Human
Error

Causes and Control

Human Error

Causes and Control

George A. Peters
Barbara J. Peters

CRC Press
Taylor & Francis Group
Boca Raton London New York

CRC Press is an imprint of the
Taylor & Francis Group, an **informa** business

A TAYLOR & FRANCIS BOOK

CRC Press
Taylor & Francis Group
6000 Broken Sound Parkway NW, Suite 300
Boca Raton, FL 33487-2742

First issued in paperback 2019

ISBN-13: 978-0-8493-8213-0 (hbk)
ISBN-13: 978-0-367-39114-0 (pbk)
Library of Congress Card Number 2005043548

Library of Congress Cataloging-in-Publication Data

Peters, George A., 1924-
    Human error : causes and control / by George A. Peters and Barbara J. Peters.
        p. cm.
    Includes bibliographical references and index.
    ISBN 0-8493-8213-0 (alk. paper)
    1. Errors--Prevention. 2. Accidents--Prevention. 3. Decision making. 4. Cognitive psychology. 5. Risk assessment. 6. Human engineering. 7. Professional ethics. I. Peters, Barbara J., 1950- II. Title.

BF623.E7P48 2006
152.4'3--dc22
        2005043548

Visit the Taylor & Francis Web site at
http://www.taylorandfrancis.com

and the CRC Press Web site at
http://www.crcpress.com

Preface

The human error problem, in general, has taken a heavy toll in property loss, human life, and the well-being of many people. Much of the toll was preventable, as may be determined by reading what is contained in this book. The basic objective of the book is to provide useful information on theories, methods, and specific techniques that could be used to control human error. Above all, it is a book of ideas, concepts, and examples from which selections or interpretations could be made to fit the needs of a particular situation. Some of the techniques may seem rather simple, basic, and rudimentary. Other techniques are fairly advanced, and a few are very sophisticated. There will be readers who welcome and beneficially utilize the contents of the book, while others may criticize and reject some concepts and techniques. This is a predictable reaction when the audience is composed of those with widely different educational levels, types of experience, and technical or professional affiliations. Hopefully, the book will be seen as a persuasive integration of useful "how to do it" information.

The scope of this book may seem rather broad, and it does have widespread applications. However, it simply applies and extends known generic principles that could help in the prevention of consumer error, worker fault, managerial mistakes, and organizational blunders. In other words, it deals with error as manifested in an increasingly complex technological society that imposes severe demands on its inhabitants. The primary focus is human error, including its identification, its probable cause, and how it can be reasonably controlled or prevented. To accomplish this, the book must be sufficiently detailed, explanatory, and practical to be of any substantive value. It is up to the reader to devise a proper framework (plan), select workable concepts and techniques (action items), and attempt implementation (need resolution).

When human error does occur, even the well intentioned may simply avoid the detailed tasks involved in attempts to control human error and let nature take its course. This is often true should they become blocked, frustrated, or puzzled as to what is actually needed, what tools to use, what could be reasonably acceptable, and what might be the likelihood of its being effective. In formulating a working model for human error prevention, there are obvious compromises that must be made of a very practical nature. This often reflects the inadequacy of the available data when and where needed and questions about the kind of analysis that would be effective at a given point in the design or operation process. The constructs presented in this book also are intended to help optimize human performance and to achieve higher safety goals. Hopefully, this book will bridge the gap and illustrate the means for achieving a fully integrated, process-compatible, comprehensive in scope, user-effective, and methodologically sound model.

Despite current efforts, the propensity for human error remains huge in many engineered products, processes, and systems. In part, this is because isolated engineering and behavioral research efforts may not be timely or entirely relevant. In addition, attempts by those in varied disciplines to simply apply human factors, design safety, quality, reliability, system safety, and behavioral information may fail because of ingrained personal beliefs, group perceptions, knowledge deficits, and limited opportunities to effect beneficial changes in design and usage. Each discipline may have commendable achievements and benefits, but human error reduction needs more direct attention and relevant remedies. Significant omissions still occur in orderly prescribed design, development, test, and management processes because of a lack of a comprehensive working model or a "how to do it" manual of good practice.

Hopefully, the contents of the book may prompt further elaboration, stimulate needed research endeavors, and provide descriptive experiences that could be emulated. At each step in our learning process, our understanding seems to reach a new higher plateau, but then we find that it is only a platform from which we are enabled to comprehend the vast amount of discoveries that lie ahead in this professional activity.

A no-cost supplement to *Human Error: Causes and Control* is available from the following Web site: http://www.crcpress.com/e_products/downloads/download.asp?cat_no=8213. Request *The Error Manual* by Peters & Peters.

While the authors of this book sincerely believe its contents will be helpful to other professionals and technicians, they cannot ensure its proper application and usefulness in a specific situation. This is because of the wide variety of people (readers of this book) with different education, training, experiences, skills, talent, subject discipline affiliations, and workplace or other opportunities to effectively implement countermeasures. Personal perceptions also greatly differ; some view human error pessimistically as an acceptable or tolerable fault that cannot or should not be appreciably changed, while others view human error more optimistically as a target needing the application of effective remedies. There are those who reject the reality of the behavioral sciences, pharmacology, or other disciplines, and such a bias would skew or distort concepts of human error and countermeasures. Despite these reservations and realities, the authors are convinced that human error reduction is a virtuous objective that can improve individual, enterprise, and social well-being. The book is dedicated to those in various disciplines who discovered the basics that are integrated into this volume and interpreted to help prevent human error.

G.A. Peters

B.J. Peters

Santa Monica, California

About the Authors

George A. Peters

George Peters is licensed as a psychologist, safety engineer, quality engineer, and lawyer by the state of California. He earned a Doctor of Jurisprudence degree after his studies in psychology and engineering. He has written many articles that have appeared in legal, medical, and engineering journals.

He has authored or edited 40 books, including nine on automotive engineering and one on warnings, and has written chapters in books such as the *Occupational Ergonomics Handbook* and *The CRC Handbook of Mechanical Engineering*. He is a current member of the Editorial Advisory Board of *Occupational Health and Safety* magazine and, in the past, has served on the editorial boards of various professional journals.

Dr. Peters has been a long-term (30-year) active member of the Society of Automotive Engineers and other organizations including as a fellow of the Human Factors and Ergonomics Society (from which he received the Kraft Award, the Lauer Award, and the Small Award); a fellow of the System Safety Society (past president and past editor of *Hazard Prevention*); a fellow of the Institution of Occupational Safety and Health (U.K.); a professional member of the American Society of Safety Engineers (former vice president, and several technical paper awards); a fellow in the American Association for the Advancement of Science; member of the American Society for Testing and Materials (a member of a number of standards formulation committees); a former fellow in the Royal Society of Health (U.K.); and a certified professional ergonomist (former member of board of directors, Board of Certification in Professional Ergonomics). He is also a former member of the board of directors of the Board of Certified Safety Professionals; a member of the American Society for Quality (former certified reliability engineer); a monthly columnist for *Quality Engineering*, for 10 years; and a member of the American Chemical Society and the American Industrial Hygiene Association.

Dr. Peters has had experience as a design engineer for large corporate enterprises and as a human factors specialist for a military agency, and has extensive experience in the analysis of management issues.

Barbara J. Peters

Barbara Peters has specialized in solving problems related to safety, human error, health, and the environment. She has taught at the university level, lectured in France and the Philippines, and has given presentations to such groups as the American Psychological Association, Georgia Institute of Technology, the Human Factors and Ergonomics Society, the System Safety Society, Cal-OSHA, and the American Society of Safety Engineers. She earned the doctor of jurisprudence degree and is a fellow of the Roscoe Pound Foundation. With her father, she has authored or edited 40 books, including *Automotive Vehicle Safety* and eight others dealing with automotive engineering. She has focused on industrial problem solving.

Contents

chapter one

Introduction

Human error is an undesirable aspect of everyday life. It is often excused on the basis that mistakes are just part of the everyday learning process. Simply put, people make mistakes, to err is human, and people are not perfect, except possibly you. In general, it is to be expected. Human error may become manifest in the form of human behavior or conduct that can be categorized as undesirable, unacceptable, careless, inattentive, forgetful, reckless, harmful, a miscommunication, human performance that is extreme in variability or beyond the limits of that expected, or an inappropriate form of risk-taking behavior. An error may be harmless, it may be detectable and correctable, it may have only nuisance value, or it may serve to predict future problems. It is certainly important if human error is a cause of 50 to 90% of all accidents, as is often concluded.

There have been efforts to mistake-proof or goof-proof various products, complex systems, industrial processes, technical and professional services, and other human interfaces. There is certainly less tolerance of human fault in the design or manufacture of mass-produced products that could have widespread or long-term adverse consequences on many people. This book attempts to better conceptualize the methodology for minimizing harmful human error.

Human error may be segregated from human fault and human failure because of customary usage. The term *human error* is a broad category that includes the clearly identifiable, easily diagnosable, generally understandable, and seemingly excusable mistakes that are very common. *Human fault* has legal connotations that include notions of the blameworthy, the punishable, and the damage-producing results of negligent or intentional error. *Human failure* includes the very significant error events that have moral consequences and may be considered nonexcusable human performance shortcomings.

As will become obvious in the book, there are many tools, techniques, procedures, remedies, methodologies, and overall plans that could be used to reduce or minimize human error propensity, variability, and consequence. They range from procedures that include a design or system safety perspective to procedures involving worker selection, training, placement, and job restrictions. The reader can select what is necessary, under the circumstances, and prepare a goal-directed integrated human error control effort.

Scenarios

The following illustrative scenarios demonstrate the various types of human error, the scope of the problems created, and the variety of remedies that could be instituted.

There are many examples of human error that pervade almost every aspect of our lives, whether recognized as such or not. In terms of *driver error*, just ask an automobile driver about the carelessness, inconsiderateness, recklessness, indifference, or self-centered driving behavior of some other vehicle drivers. It is the "other driver" that may cut into a small space between vehicles without first signaling a lane change. Other drivers may travel well in excess of the posted speed limit, thus violating the rules of the road. Some drivers may fail to stop at a stop sign or for a traffic control device displaying a red-light signal. At certain locations, there may be black tire marks on pavement curbs or on the physical barriers separating the direction of vehicular travel on a high-speed highway. When a driver wanders from lane to lane, steers curves widely, or does not act appropriately, it may be an error in terms of the social norms that have evolved for a complex societal situation. Such driver error is fairly well tolerated and only rarely leads to significant accidents. The driver at fault may be acting out, may be impaired naturally or by drugs, or the driver errors may be the manifestation of a conduct disorder. The driver error may have been induced by the highway traffic control system, by deficiencies in vehicle design, or by the complexities of the social interactions of drivers. To reduce this form of human error, the focus has been on police enforcement of the rules, traffic court penalties for nonconformance, and constantly evolving highway improvements for the continual increases in the number and type of vehicles on the road. Driver error is often accompanied by the obscuration given by the rationalizations that generally accompany human error and possible personal fault.

Even with devoted *maternal child care*, fatal human errors can occur. One mother was closely monitoring her 2-year-old child when the telephone rang. The child was in a fenced-in area of the backyard, so the mother ran to answer the telephone in the kitchen. When she quickly returned, the child was not in sight. Then she saw her child at the bottom of the swimming pool. To get to the pool, the child had to have climbed over a cinder block wall or had to have unlatched a gate. It was her only child that drowned, and she quickly concluded that she made an error in relying upon the security of the backyard enclosure. This belief of a maternal human error was accompanied by a deep sense of personal guilt and a lasting depression. The father could not handle the consequences of the human error and secured a divorce. There were many child-resistant gate latches, enclosure fences, and pool covers. There were many types of alarms because of the frequency of such accidents. As the mother said, "My attention was diverted for only a few seconds," but such human errors must be logically justified for personal emotional recovery. Human error prevention or compensatory measures can be lifesavers.

A young 18-year-old male student wandered over to the ocean-side park near his home. He had developed the habit of watching the beautiful sunset while being alone to think about and resolve his daily experiences. This day he sat on the edge of the familiar cliff, more than a hundred feet above the rocky shore below. Suddenly, the cliff face collapsed and he fell to his death on the rocks below. He had not noticed a small stream nearby that had eroded and weakened the spot he chose to sit upon. He had relied upon the municipality's expert park service to ensure the integrity of the sight-seeing spots or to fence them off and warn of any danger. His absence of self-reliance prompted the apparent risk-taking behavior that others could not understand. It was alleged to be a human error of judgment, since others asserted that they would never sit on the edge of an earth cliff a hundred feet above a bed of sharp rocks.

About midnight, two couples decided to go to an apartment complex for a dip in a hot tub and a swim in the pool. There had been some moderate drinking and there was a sense of pleasurable excitement among these four individuals aged 19 to 23. The older male suddenly ran from the hot tub and dove into the pool. The younger male quickly followed suit, but he misread a half-submerged depth marker on the side of the pool as 8 feet instead of 3 feet. When he dove into the shallow end of the pool, he struck the bottom with his head and was rendered a paraplegic. In the excitement of the evening he made a human error that changed his life forever. It is well known that people act differently in a close personal relationship where exhibitionism before others may be an impulsive risk-taking behavior. This type of diver fault might have been prevented by a limited access to the pool during the late evening hours, an appropriate supervision by lifeguards, a better indicator of pool depth, or an after-hours pool cover that prevented untimely use.

An unskilled helper was assigned to clean some printing machines while they were operating. He was given a verbal caution to be careful and shown how to clean the rotating cylinders. While doing this, the cleaning rag he was holding suddenly got caught between two in-running rollers (a nip point). The rag was pulled in so quickly that he did not have time even to release his grip. His hand was crushed between the two rollers and there was a degloving injury (the skin removed like a glove). The plant manager investigated and concluded that the helper was careless, inexperienced, and at fault for his own injury. Such victim fault opinions are quickly, easily, and summarily concluded on the basis of an assumed human error. The absence of guarding, automatic shut-offs, or work procedures requiring jog-only (limited movement) cleaning operations could place some of the fault on the work supervisors or shop owners and managers.

A machine operator of a vertical plastic blow-molding machine had been told to watch the parisol (plastic extrusion tube) for anomalies (anything that looked wrong). If he saw anything wrong he was to reach over the top of the barrier guard to the parisol and "whack it off" quickly. One day he did it, but his hand got caught between two steel plates and he lost some fingers. Was it his human error, his work practice instructor's error, or the error in designing and installing a barrier guard with ostensible safety features possessing the capability of reach-over access to a danger zone?

A wholesale grocery delivery truck had a side door into a refrigerated storage area for perishables. There was no grab handle for the truck driver who had to unload the cargo. One day he put his foot in a movable stirrup and his hand on the side of the door entrance, lifted up, then stepped inside quickly to regain his balance. The refrigerator floor was icy and he slipped and suffered a back injury. He was told he was at fault because he was not careful enough. Subsequently, the trucks were equipped with long grab handles reachable both from ground level and from inside the refrigerated section. The step (stirrup) was made rigid, larger, sturdier, and nonslip. The refrigerator floors were made nonslip. His employment records continued to indicate a human error, and he was laid off at the next employment cutback.

There are many forms of medical error that can jeopardize the health and safety of patients. It may be in the improper processing of prescriptions, surgical mistakes, or user error in dealing with medical devices (misuse). In the U.S., a 1999 report by the Institute of Medicine indicated that hospital medical errors killed between 44,000 and 98,000 people each year. In Minnesota, a report released in 2005 on 139 hospitals indicated that there were 13 surgical operations on the wrong body part, 31 cases of foreign objects left inside surgical patients, and 21 preventable deaths in 2004 (Davies, 2005). This included five fatal medication errors, four instances of a fatal misuse of medical devices, and a cluster of mistakes involving surgery on the wrong vertebrae. Because of the medical device design flaws, the U.S. Congress enacted the Safe Medical Devices Act, which required good manufacturing practices (GMP). One report about patient-controlled analgesia (PCA) pumps concluded that a human factors improvement of the pump interface resulted in a 55% reduction in user errors (Murff, 2004). In terms of operating room alarms, in one study they were correctly identified less than half the time. But alarm detectability was 93% accurate with a better designed alarm system. Is it error when a physician initially guesses at the correct patient medication and dosage, knowing that he can observe the objective and subjective reactions, then modify the dosage or change the medication as necessary (an iterative narrowing approach)? What if the patient does not return for a checkup? There is continuing improvement of the procedural checklists for surgical interventions, usability analyses, observational techniques, and the systematic evaluation of devices, diagnostic procedures, and therapeutic regimens.

The trial of a lawsuit ended on a Friday afternoon. The jury recessed to discuss the case and reach a verdict. After a short time, one middle-aged juror decided to vote for the defense to cut short the deliberations because her husband was about to arrive at home and would want a hot meal prepared for him (according to a subsequent written declaration). Previously she had been leaning in favor of the 80-year-old female pedestrian in a marked crosswalk who had been struck by a 22-year-old male driver who failed to stop in accordance with the rules of safe driving. The juror's human error was recanted, but could result in the burden of another trial before a jury. Similarly, trial judges frequently make mistakes about the applicable law, and if significant, they are termed reversible error by an appellate court.

A semiskilled supervisor walked over to a horizontal plastic molding machine that operated by the use of a vacuum process to draw a thermo-setting plastic into the shape of a tray or container. He had been told that there was a misalignment of the sheet plastic feeding into the mold. The mold area was enclosed with a see-through barrier guard intended to prevent anyone from reaching into a danger area (the opening and closing of the mold) in order to make adjustments while the machine was still running. However, there was a hole that he knew he could reach through without interrupting the flow of production. He reached in, made an error in timing and finger placement, and lost several fingers. In addition to the human error of bad timing and insufficient motor coordination (dexterity and space location), he manifested poor human judgment in not stopping the machine. He previously failed to plug a known hole in the physical barrier or guard that was interposed to prevent anyone from reaching into the danger area of the reciprocating mold sections. He made an error in believing he was more macho or had superior safety skills than the normal unskilled machine operator. Since production was king, human error was foreseeable. But no harm would have resulted if the interposed barrier had no loopholes. The machine should have been easily stopped for adjustment by a convenient emergency stop button, without overheating the plastic, otherwise damaging the product, or seriously interfering with the production of the plastic trays. The machine could have been better designed to forestall human errors by both machine operators and work supervisors.

Food safety mistakes (human error) may occur in the use of certain foods (diets), in their formulation (recipes), or in subsequent changes (the substitution or addition of various ingredients). A food intolerance may produce various symptoms such as nausea, cramps, bloating, gas pains, flatulence, diarrhea, or borboygmi (gas rumblings in the intestine). This discomfort may be from sources such as milk, yogurt, sour cream, ice cream, and many other processed foods. The digestive problem may be from an intolerance to milk sugar. Such lactose intolerance may effect ethnic groups differently, such as 30% of Americans and 90% of those of Chinese ancestry. The intolerance is from an insufficiency of the enzyme lactase that acts on milk sugar (glucose and galactose) so that it can be absorbed into the bloodstream from the small intestine (Ryan, 2004). The food user must cope by avoiding foods that list lactose as an ingredient. But food labels may be incomplete or in error (by intention or mistake). Self-education as to which foods contain lactose is necessary to help avoid errors. Lactose-reduced foods are now available. Lactase enzyme tablets may be used when there is poor control over a meal prepared by others, served in a restaurant, or where there may be lactose suspected in the food purchased. Thus, the potential for food safety errors is present in many ethnic cultures by consumers, chefs, grocery stores, and food producers.

Building flaws may result from the error of architects or earthquake engineers who rely upon the Uniform Building Code and seek approval of their plans from a local government building department. The Imperial Valley earthquake of October 15, 1979, resulted in significant damage to a relatively new county services building. The cause was the unsymmetrical placement of the walls and irregular open spaces without walls. An article in a scientific journal (R.A.K., 1980) stated that the building met or exceeded all seismic safety requirements in the applicable building code, but the purpose of the code was to protect lives, not to minimize the damage. General building codes are not enough and have unavoidable limitations in the design of buildings. It may be easier to rely on generally accepted codes, but human error can be reduced only by depending on the human judgment of a building engineer and the functional or financial commitment of the building owner. It requires proper calculation of strength, layout, durability, future building improvements, wind loads, assumption of the primary duty of earthquake resistance in accordance with foreseeable circumstances, and predictable human usage patterns. If earthquake susceptibility, from an engineer's undercalculation error, becomes known to the building owner many years after the design and construction is completed, there may be a legal obligation for a warning to the occupants or for the owner to undertake a retrofit action. This is particularly true for unreinforced masonry buildings, which in one earthquake resulted in 3000 buildings being designated as unsafe by the building inspectors (USGS and SCEC, 1994). Errors by architects, engineers, and building owners can cause foreseeable personal injury, property damage, and business interruption. Adequate knowledge to supplement existing building codes is an essential preventive measure. Corrective measures may be an imperative necessity.

Chemical engineers often deal with the processing and storage of dangerous chemicals. At one pesticide plant, there were three storage tanks holding about 40 tons of liquid methyl isocyanate (MIC). This was an ingredient used in the manufacture of carbaryl (the pesticide better known as Sevin). In the early morning of December 8, 1984, a storage tank leak of the toxic MIC gas formed a cloud over the city of Bhopal, India, that killed between 3800 and 15,000 people, caused permanent disability to more than 2700 people, and otherwise injured an estimated 300,000 people (www.bhopal.net). The frequency and severity of injury, in such cases, are only estimates by various entities having different interests, such as government agencies seeking to correct wrongs while still trying to achieve the restoration of jurisdictional employment, local economics, and the appearance of hospitality to business enterprises. Also having special interests are the companies that might bear some fault and liability, the survivors' groups seeking compensation, the victims' advocacy groups seeking redress, community groups seeking restitution and justice, and the personal perceptions of journalists who report on such disasters.

The apparent cause of the Bhopal tragedy involved multiple human mistakes (errors) and human acts of neglect, omission, and misunderstanding. A leak was discovered in one of the storage tanks, but a work supervisor assumed it was just a water leak and temporarily postponed action (a mistake). Earlier cleaning procedures left water in the tank (a mistake), resulting in an exothermic reaction (a chemical reaction liberating energy in the form of heat). When tank pressure increased, a valve burst, the MIC leaked, and it vaporized into a toxic cloud. Questions arose about worker training (the cleaning process), disrepair (the valve failure), and sabotage (an intentional act).

The preventive systems failed; this included a spare holding tank for leaking MIC (it was not empty), a refrigeration system to cool and inhibit exothermic reactions in the storage tanks (it was inoperative, apparently shut down to save costs), a water curtain that was ineffective (it did not reach high enough), a caustic soda detoxifier that had been turned off, a flare tower that had been disabled (a pipe removed for maintenance), some pressure and temperature instruments that were alleged to have been unreliable, and a shortage of redundancy in terms of computer backups and automatic shutoffs (Pizzi, 2004). There had been at least three prior small accidents at the plant involving MIC. There was no community emergency plan for the city of 800,000 residents, such as safety instructions and defined evacuation routes should there be a major chemical accident. Well before the accident, the local government had attempted to get the company to move its plant from the congested city to a safer rural area.

A financial settlement between the Indian government and the plant owners was reached in 1989 for U.S.$470 million, but it was a real public relations disaster. Comparisons were made with the U.S.$5 billion punitive damages paid for the *Exxon Valdez* oil spill in Alaska. In 1985, a year after the Bhopal accident, a pesticide plant in West Virginia had a leak of chemical aldicarb oxime that injured 135 people. About that time the chemical industry trade associations, first in Canada, then in the U.S., adopted a voluntary responsible care program to more openly provide chemical health and safety information and to better define desirable conduct relative to chemical transportation, toxic emissions, and pollution prevention.

Precautions taken in name only for human safety or in a form that can be easily neglected may not be effective under many circumstances that could be predicted given a diligent and competent inquiry. Precautions require appropriate attention during initial design, but also during continued surveillance of in-plant chemical plant operations, plant shutdown, maintenance, repair, improvement, conversion, expansion, abandonment, dismantlement, and ultimate disposal. Just as most safety problems involve multiple causation, they usually require multiple preventive remedies. If human conduct (error) problems predominate in accidents, is there proportional in-depth attention given to the analysis and prevention of human error?

Caveats

Importance

Human error infests almost every aspect of human life. It is so common that it is easily excused, rationalized, and then forgotten. However, human error is an important cause or contributor to accidental property damage, significant personal injury, and avoidable human death.

Common sense

Almost everyone believes that he or she knows a lot about human error and its prevention. This every-person knowledge may result in quick, simplistic, impulsive, and emotionally charged conclusions. The knowledge may be completely subjective in nature, superficial in character, and general in application. It is often wrong in terms of an effective remedy. Human error and its prevention are just too important to leave to common sense, however well motivated.

Sloganeering

There are slogans almost everywhere that reflect common misunderstandings about human error. There may be signs or statements to "be careful" and to "avoid mistakes" that assume a general level of indifferent carelessness, that people do not try to avoid mistakes, and that errors can be voluntarily reduced with a little effort. However, simplistic slogans may be perceived as inapplicable, may be ignored, are quickly forgotten, and can be counterproductive. They suggest that there is little real concern or understanding of the complexity of human error, its causes, and its control.

Bias

There are many scenarios in which human error is deemed the culprit in the absence of detailed knowledge. It may be assigned the blame without a comprehensive analysis. Human error may be an easy target in unsophisticated accident reconstructions. It may be more difficult to blame equipment, training, or other factors such as any contributing precursor events. There is often a lack of situational awareness, that is, what was going on around an incident and how that affected what is real and what is not. In essence, there is often a bias in terms of easy targeting of what or who to blame and how to explain it away.

Error proofing

There is a pervasive belief that modern technology cannot be foolproofed and operations goof-proofed. Such concepts assume that errors are made by fools when, in fact, they are an equal-opportunity situation that does not discriminate against anyone. The concept of goof-proofing assumes a perfect remedy and usually infers a simple solution. Even the best of us can goof. It is the characterization of the people who commit errors or goofs that is improper. The objective is better achieved by a rational study and understanding of all of the factors that may contribute to error. In fact, the remedies need only be reasonable under the circumstances, which are often quite complex.

Need

In general, a much more sophisticated and productive approach is needed for human error identification, cause, prevention, and mitigation of consequences. It is obvious that there has been considerable historical experience in dealing with human error, but the results have been insufficient and inadequate. Something more logical, realistic, detailed, scientifically based, and multidisciplinary may provide the missing details, concepts, and techniques that are urgently needed.

chapter two

Data collection

Introduction

The initial approach to any human error inquiry is appropriate data collection. This is true whether the focus of the effort is on just one highlighted human error problem with the objective of improving human performance or a more ambiguous and broader scope effort to specifically identify and effectively resolve all the potential sources of human error in a product, process, system, or service. It may be a postmarketing or after-the-fact attempt at *corrective* action, where the problem is known but the real causation and remedy are at issue. It may be an early design situation involving a proactive or *preventive* action effort where potential human error problems are not known and ambiguity may seem to prevail since no actual adverse or reportable event has as yet occurred. Thus, the data gathering may be comparatively narrow in scope or somewhat comprehensive in character. Despite the focus of the effort, relevant data can and should be the objective building blocks that can form a firm foundation for opinions of value and conclusions that have real substantial merit.

Evidence-based data

Subjective judgments are often made and used in determining probabilistic human error estimates. Such estimates may be supported by peer consensus, prior usage, common beliefs, or from the absence of contrary data. Too often it is from poorly formulated, biased, or extraneous test, customer adjustment, warranty, or field information systems. If of low value, the data can lead to poor choices and erroneous expectations. Since persistent or continued use of bad data may promote self-deception and false confidence, some reality check or reference to an external criterion may be desirable. Evidence-based explanations are more credible and have greater potential value. Stated another way, objective data are preferable to subjective data. If highly subjective estimates must be made, this fact should be disclosed to prevent unjustified applications. Serious attempts should be made to improve the foundation and quality of the information that may be provided to others.

The human error estimates may be made in the form of just a few categories. Each error might be classified as high, medium, low, or improbable in terms of frequency of occurrence or the severity of consequences. Such a crude classification system encourages uniformed guesses, lacks precision, and provides little useful descriptive content. Assuming that the devil is in the details, the need may be for greater depth for better understanding. Refinement of the data may result in greater qualitative sophistication and potential usefulness.

Sources of data

There may be some immediately available human error data that can be considered, reviewed, analyzed, or further investigated for deeper meaning (that is, to determine actual causation and the degree of its veracity). This might include data that are categorized as reliability, system safety, human factors, project or peer group, field service, dealer adjustment, or even the industry folklore. The purpose of the self-check of available data is to discount possibly misleading or inquiry-ending information. The kind of information that may be discounted includes subjective personal opinions based on rather quick assumptions, long-held unfounded commonsense beliefs, the self-protective excuses of work supervisors or sales personnel, the statements of the obviously ill-informed, and the conclusions of those apparently biased in some respect.

Ideally, there might be a portfolio of potential human error problems. Perhaps a human error guide should be in constant preparation (similar to a design manual or a lessons-learned data bank). There may be a rich source of data from accident reconstruction reports, loss control data, or in-service repair compilations or summaries. Trade standards, directly or indirectly, may suggest industry-wide problems that might include human error. Prior attempts to correct might be revealing. Similarly, government regulations and standards may include requirements intended to prevent human error of a particular type. Professional society literature may be a valuable source of introductory data, remedies, and guidance principles. The objective of this phase of a human error inquiry is to acquire a preliminary set of data and a general familiarization with the current state of the art.

Often overlooked are *historical documents*, generally design feasibility and test reports. They may be stored or hidden in various company archives or facilities. They may lead to failure reports, malfunction reports, trouble reports, design evaluations, customer complaints, schedule delay reports, quality control records, and injury compensation decisions. The task is to extract what might be of value from these static sources of information.

In terms of further data generation, an *interview* or personal inquiry technique should reveal new useful information. The interview may be with project personnel, shop workers, in-field technicians, or users. The objective is to listen and learn, not to advise or teach. It may help in understanding perturbations, confounding, and appropriate cleansing of data, both present and past. It would be wise to consider the effects of moral, ethical, legal, job retention, and customer satisfaction issues on the type of human error being discussed. It may prove Murphy's law that anything that can go wrong will go wrong (in terms of human error). There may be an implied theorem that to err is human, but it is easily checked (thwarted). That is, personal interviews should prompt open disclosures and serve to reveal the antidotes.

Perhaps the most useful data are those gathered using direct *observation* of humans engaged in the use of a product (usability), its selection (appropriateness to need or desire), its installation (unpacking, assembly, and break-in), its servicing (maintenance, repair, and adjustments), and its disposal (inerting, disassembly, packaging, transportation, destruction, and recycling). Observation of shop personnel in manufacturing or test operations may reveal problems that can be discussed in group sessions or team meetings to gain a better understanding from those directly involved. Observational studies may benefit from videotape recordings that can be played back in slow motion. Good data may be generated during task or job exercises in demonstration or validation facilities or in simulation or usability laboratories. It is particularly helpful to identify human error in carefully controlled situations.

Interesting data may be gathered from *experimental research* on narrow issues of major importance that justify the costs and time delays. The preliminary context of such research studies may be found in scientific and engineering journals, relevant books and reports in the public domain, and by Internet searches. *Focus groups* and consumer clinics may provide information translatable to useful data. Independent laboratories may provide interesting data. An *iterative design and testing* process suggests that human error data accumulation is an ongoing or continuous process. *Violations of code* may suggest that there could be unaccountable general error variances conducive to human error and the weakening of human error controls.

It is not advisable to ignore human error just because it may occur rather infrequently, particularly if a disastrous consequence could occur. Errors believed to be remote are too often the unexpected causes of problems, particularly in cumulative or joint interactions. The capability of neutralizing, inerting, or foiling such an infrequent error may be simple, of low cost, and easily attained. Error frequency and severity classifications are often treated as being relatively independent of the complexity, effectiveness, and cost of other applicable remedies. In essence, gather all human error data at this stage of inquiry. Assessment of its relative importance may be left to subsequent evaluation, assessment, prioritizing, culling, and risk acceptability criteria.

Clarification and purification

It may be helpful to consider the differences between raw data, relevant data, and productive data.

A recent analysis of chemical plant problems, defects, and process upsets indicated that there are five main sources of these problems (Price, 2005):

Maintenance practices	18%
Maintenance materials	7%
Raw materials	5%
Design	5%
Operational discipline	45%

The major human error problem was called operational discipline, that is, the failure to follow procedures and to stay within the operational envelope. The second human factors problem was maintenance practices, that is, the inspections, repairs, and modifications of equipment. Together, these two human error categories account for 63% of the chemical plant problems.

Looking more closely, many questions arise. Does the failure to follow procedures mean a failure to follow *customary* practices that should be used, a failure to follow *oral* instructions that should have been given by a supervisor or trainer, or a failure to follow the prescribed procedures in a *written* manual? If written procedures are available, are they too difficult compared to the literacy of the user population, in terms of school grade levels as tested by the Flesch-Kincaid, Coleman-Liau, Bormuth, or other readability tests? Very often, mechanics do not consult procedures in printed service manuals or those that are available on computer terminals, since they often use the diagnostic challenge of trial-and-error methods to solve problems. In some situations, written instructions are not conveniently available, practical, reasonable, or understandable. It is the specifics that can yield useful information about the failure to follow procedures.

These chemical plant classifications are broadly inclusive and so general that they merely point to general problem areas that need investigation. A focus on general categories often produces quick general remedies, when specific targets and tailored countermeasures would be more effective. In such situations, the underlying data are often confounded, biased, confusing, speculative, and a mixed bag of excuses, rationalizations, and brief conclusions. Such information can only be considered *raw data*, which needs clarification, better definition, refinement, and purification.

When the data are cleansed, elaborated, interpreted, and properly selected, they may have the appearance of *relevant data*, which is more directly useful. While relevant data better describe a human error problem, the underlying cause may still be somewhat obscure. There may be hasty attempts to institute quick fixes, or there may be well-reasoned grounds for more precise solutions. Some examples may be instructive.

Medication errors have occurred at a 2% rate of admissions in some hospitals. More specifically, these are human errors resulting in adverse drug events to hospital inpatients. This is still a general classification since there can be many specific causes of such errors relevant to patient safety. The general remedy was to install digital computer systems that require nurses to scan the patient's identification wristband, the bar code on the prescription containers, and the patient's code, and to insert the nurse's code. This helps to ensure that the medication dispensed is correct for that patient, given at the right dosage, and administered at the right time. The cost is high for a general preventive program.

A similar problem has existed with the correct dosage and timing of pharmaceuticals for outpatient consumers. The human error may be defined as noncompliance with the prescribed regimens. One general solution is the use of unit–dose packaging, for solid and nonsolid medications. It is often called convenience packaging rather than compliance packaging. The specific causes of noncompliance may or may not defeat unit-of-use or unit–dose delivery systems, such as tablets on blister cards with instructions on the back and unit–dose vials for liquids, gels, lotions, and creams (for topical and oral dosing). Another approach is the use of reduced space bar code symbologies and radio frequency product identification tags for high-speed readability, with traceability software, that can provide a health care delivery system with less opportunity for human error.

Radio frequency identification (RFID) labels and tags can hold a great deal of information, such as drug expiration dates, product numbers, serial numbers, lot codes, user codes, location codes for each stop in the supply chain, and tracking data. Such information could be invaluable for human error investigations.

Such systems are sold on a cost–benefit basis, so the need or demand must be quantified to some degree as relevant data become available, suggesting a potential benefit as the relevant data become available. The offsetting costs are based on economics, technical feasibility, effectiveness, and resistance to change. At one time, there was serious resistance to child-resistant packaging and unit–dose packaging even for drugs that could harm children, so resistance as well as technical feasibility are time-dependent variables.

Hospital medical errors include surgical mistakes such as operations on the wrong body part, operations on the wrong patient, wrong procedure performed, and foreign objects left in the patient after surgery. A variety of countermeasures have been employed, such as appointing a quality oversight committee for surgical procedures rather than marking the skin (which may shift in overweight patients). It has been found that what works in one hospital may not work in another because of staff, professional, cultural, and economic considerations.

To help correct intensive care errors, continuous two-way visual patient observation and audio communication has been achieved by use of television cameras and microphones at various locations (telemedicine is attractive). An emergency physician or medical specialist can see a patient, talk to the patient, examine current diagnostic indicators, and review the patient's medical history on a television screen at a remote location.

In essence, relevant data, whether general or specific, can be very helpful in determining general countermeasures. But sometimes the data are just not enough to solve other problems. A relevant human error problem may be known, but hard to understand, and it seems impossible to define a real remedy. Stated another way, the error, the cause, and remedy may seem abtruse, inexplicable, and a puzzle that cannot be solved. For example, a design group in the polymer industry identified a human error problem as critical and assigned it to a senior technical specialist. That specialist spent several years, part-time, in a frustrating and unsuccessful effort to resolve the problem. Similarly, in another company, a technical committee was established to better define and solve a significant human error problem. They could not solve the problem because, as one participant claimed, they allowed groupthink to retard their reasoning process. They had ample data, had research performed, but the problem remained enigmatic. The information they had was relevant to problem solving but was not enough. Such misfortunes are one reason for this book.

Productive data are those that are pregnant and ready to give birth to countermeasures. The are a form of decoded raw data. They are immediately useful and capable of inferring a remedy. They are usually the result of diligent and painstaking effort by a specialist with the correct personal perspective, a zealous form of inquiry, an appropriate attitude, and the kind of specialized knowledge described in this book.

Even highly refined productive data need a follow-up after implementation of the obvious countermeasures, both to determine the real-life effectiveness of the remedy and to capture residual errors, previously undisclosed associated errors, and possible new side effects. Errors that may have seemed incidental and harmless may be revealed as serious risks during the follow-up. Humans are innovative and can circumvent inconvenient or misunderstood error control efforts, so follow-up is important and could result in modifications of the error countermeasures. It is a form of useful data collection.

In the search for refined productive data, there should be a conscious effort to exclude bad data that could contaminate the process. In reviewing information for possible human errors, what criteria are used to select good error data or characterize it as bad data? There may be *a priori* or presumptive bias in the selection of human error data because of what is expected, desired, or could prove or disprove a contested point. There should have been independent decisions and caution in accepting anything from those with a vested interest in obscuration, hiding the ball, or maintaining the status quo. The errors should be proximate to an identifiable harm, not so remote as to be meaningless. There is always a danger in empirical or *a posteriori* associations (going from effect to cause), so some substantiation is important. If bad data are suspected, an uncertainty analysis or estimate can be made to help in making correct decisions and reaching valid conclusions.

In the chapters that follow, there is a great deal of information that could stimulate data collection efforts of value.

Cautions

If the data that drive conclusions are an important preevaluation event, there are some important cautions that should be observed:

1. Do not assume that all the *data being collected are clean.* There is always some inaccuracy, contamination, and confounding of the data collected from an uncontrolled real-life situation. Are the data adequate for the purposes? There will be subsequent evaluations to determine the quality, validity, and usefulness of each type of human error in the accumulated data bank.
2. Do not strive for *perfection.* Increasingly higher accuracy levels generally require exponentially higher costs, and this may not be justifiable. Are the data equal to the task? If refinement is needed, there are techniques, described in this book, that can be utilized in the subsequent phases of the human error reduction effort.
3. Become a *professional listener* in interview sessions. Carefully select those to be interviewed and treat them with dignity and respect. Do not coach those being interviewed. Do not infer some accountability for what they may reveal. An air of confidentiality and a joint effort to identify and resolve problems are the necessary themes. Maintain the focus on human error, after an initial accommodation of egos and perceived roles. Mutual confidence is the hallmark of a productive interview relationship. Subsequent analysis is generally helpful.
4. Be careful to maintain *unobtrusive observation.* In the presence of others, those observed will try harder and use modified procedures. For example, a maintenance technician who hardly ever looks at a service or repair manual, preferring a trial-and-error approach, may do what he believes is expected and follow the step-by-step instructions in the manual. His behavior is not natural, it is modified.
5. Learn to deal with *difficult sources.* There are always some shortcomings in the production of data. Your initial contacts may or may not have good reason to protect the data on which they could have expended considerable time and effort. In fact, you may be perceived as the one at fault for being overly aggressive in your data demands. Attempt to better understand others and develop a trusting relationship with mutual rapport and rewards.
6. Exercise due care, a circumspect demeanor, a respectful diligence, and an overt friendliness when contacting and dealing with employees having a *different management* supervision and those outside the company, such as a user population in a particular culture. Those who participate in human error studies are generally cognizant and

sensitive to the delicate nature of the personal matters involved in such studies.

7. While direct personal contact and professional analysis are required, use *technology* to augment the process, ensure adequate memory, provide appropriate data processing capability, permit economic statistical design, and yield graphics that can effectively communicate.

8. Utilize *all sources* of raw or transferable human error data, including those available from commercial and government sources (such as injury and fatality compilations), those that can be extracted from peer professional reports (such as incident and near-miss data), those that can be considered customer feedback (such as customer complaints), as well as readily available data and those that need to be generated for your specific analyses. There is a strength, in both numbers and a variety of sources, for your raw human error data to be used for predictive purposes.

Caveats

Foundation

The data collected will serve as the bedrock or foundation for further analyses. Thus, they should be good, reliable, and useful. But data refinement is a continuous process. What seem like good data at an early date may undergo astonishing changes with further in-depth study, testing, checking, verification, and concept elaboration. It is often an iterative process to achieve the necessary clarification and simplicity.

Enrichment

Human error occurs within some context, a set of circumstances, or situational surround. To fully understand an error, its causes, and its prevention, there is a need for supplemental information. There is value in knowing precursor circumstances, the chain of events, what prompted or triggered the error, any biases in the descriptions of what happened, and the prior history of similar incidents. Contextual enrichment may seem to add to complexity, but it actually serves to help in understanding, simplification, and implementing effective remedies.

No data

There may be situations for which immediate corrective action is mandated, despite the lack of appropriate data or any data at all on the topic. Subsequent chapters in this book suggest the procedures that could be used in such situations. Those analyses and actions are only interim measures that require further data collection, verification, relevancy checking, correlate exclusion, and effectiveness evaluations.

Accommodation

The data collected should be sufficiently robust that they include different perspectives and concepts as to what constitutes acceptable and relevant human error data. The central tendency (consensus) or truth of the matter may emerge later in the debate, learning process, uncertainty reduction, or information-checking process.

chapter three

Risk assessment

Introduction

After the raw, relevant, and productive data on human error have been accumulated, the data should be processed to determine their potential value relative to a chosen risk goal. The targeted goal should be reasonable under the circumstances and achievable with the resources and time available. It may not be possible to have an analysis and evaluation process that is all things to all people. The process may not conform to some standardized procedures. It does need to be effective in terms of the needs and demands unique to a given situation.

The road map that follows is illustrative of a detailed, organized, and systematic approach that could attain real, meaningful results. It is expected that flexible adaptations of the road map will occur, but the fundamental methodology should be retained wherever possible. The basic philosophy includes worst-first (*pareto*) prioritization, simple solution recognition and implementation, continued iterative analysis, objective fact finding, provisional and progressive remedies, confirmatory test validation, and conformance with social responsibility criteria.

Definitions

Some basic definitions that would be helpful in understanding risk analysis include:

1. A *hazard* is that error capable of causing harm. It is a potential source of harm. Harm is defined as injury or damage to people, property, or the environment.
2. A *risk* is the result of error frequency combined with or multiplied by the severity of the consequences resulting from its occurrence. It is the probability of harm and the severity of that harm. Error frequency may be the resultant of the exposure duration, the opportunities for error during that given exposure, and the probabilities of detection and avoidance of harm during that exposure.

3. A *danger* or unacceptable risk is that which exceeds some given acceptance level. The level may be excessive preventable danger (a legal risk–benefit criterion), a risk not in conformance with or beyond reasonable consumer expectations, or a risk beyond a defined acceptable level of risk (ALOR).

4. A *control* is that which reduces the amount of risk or its statistical variance (variability).

5. *Accuracy* is predictive functional certainty.

6. *Cause* pertains to underlying human conditions that result in human error, including but not limited to psychological factors, skills, habits, stress, knowledge, brain function, pathological conditions, physical attributes, and human information processing capability. Cause may be direct, indirect, inferential, or circumstantial.

7. *Risk factors* are those conditions that are associated, correlated, or somehow coexist to some degree with a defined resultant human error.

8. *Risk criterion* (standard) has many definitions, each of which furnishes a different overall program objective in terms of risk reduction. It also creates a different personal perspective or meaning given to various errors, fact patterns, and remedies. One or more of the following types of risks (controlling criteria or overt standards) should be identified to give direction to the risk assessment and error evaluation effort:

   • Personal injury risks
   • Property damage risks
   • Corporate or mission accomplishment risks
   • The risk of accidents in general (however minor)
   • The risk of adverse incidents that could require the time, effort, cost, or attention diversion of the risk evaluation entity
   • The risk of a loss of customer satisfaction, including user expectations
   • The risk of creating system integration problems, including assembly, fit, material handling, transportation, and reliability interactions
   • Risk management risks, including insurability, exclusions, all risk coverages, and premium amount
   • Competitive marketplace risks
   • Public opinion risks
   • Contract compliance risks, both domestic and international
   • Civil liability risks, including product liability and business interruption
   • Regulatory compliance risks, both domestic and international
   • Risk of credibility loss within the organization, by the client, or by an auditor
   • Corporate malfeasance risks, including corporate manslaughter (in countries where applicable)

Note: What is deemed important by one group of evaluators may not be as important to another, given different circumstances, motivations, and skill orientations.

Classification models

The first step is an attempt to realistically perform a preliminary screening of the available data. It is assumed that the data are somewhat refined, and not consisting of just very broad, general, and superficial categories. It assumes that an attempt has been made to determine meaningful cause, such as a root cause analysis or a professional human factors error classification. The purpose is to screen out unimportant data clutter. What is eliminated are errors that could not have an adverse consequence or are highly improbable of occurring. What is retained are data having some relevance to a selected risk criterion and some materiality in terms of probable consequences. A careful screening process will err on the side of inclusion at this step in the process.

There are a number of approaches that can be taken to classify human errors as to risk and undesirable human performance. The classification procedure that is often best suited is that which is analogous to that used in conventional failure mode and effects analysis (reliability) and fault tree analysis (system safety). This avoids unnecessary duplication, time, and costs. It facilitates technical communication, multidisciplinary cooperation, and productive interaction. Reliability engineering has emphasized parts failures (failure rates) and parts or equipment configurations that enhance long-term service life and intended function. System safety has tended to emphasize hazards and risks to physical equipment and software. Human error analysis is more directly related to the human side of equipment and systems, that is, the prevention of unwanted human performance. There is no question as to the need for human error specialists performing their unique analyses to achieve meaningful results.

Human error is evaluated by assessing the frequency of each type of human error, the severity of consequences, and the degree of risk that results. However, this is only a first step in preventing human error or controlling its adverse effects. It is important to recognize that the words used by some specialists may sound similar, but the meaning may be quite different, because of the individual perceptions, the knowledge base, and the objectives of each group of specialists.

The next station on the road map involves an evaluation and classification of errors.

Frequency estimates

It is instructive to look at system safety engineering analyses. Initially, system safety estimates were quantitative in character. The attempted accuracy was 10^{-6} (that is, one in a million). This level of precision could not be obtained for all estimates, so analysts used what they had in terms of a mixture of precise and imprecise estimates. This gradually changed so that broad classifications were being used, similar to those in Table 3.1.

Some analysts have narrowed this to only two categories: the possible and the impossible. The fewer categories, the less meaning and effectiveness that might be given to the process. For human error, broad categories suggest

Table 3.1 Qualitative Error Frequency Classifications

1	NA	NA	Always
2	Frequent	Frequent	Frequent
3	Likely	Reasonably probable	Probable
4	Occasional	Occasional	Occasional
5	Unlikely	Remote but possible	Remote
6	NA	Extremely improbable	Virtually never
7	NA	Physically impossible	NA

Source: Adapted from Ericson, C.A., *J. Syst. Safety*, 6–9, 2004; Firenze, R.J., *Professional Safety*, January 2005, pp. 28–35; Eisner, L., Brown, R.M., and Modi, D., in *Compliance Engineering, 2005 Annual Reference Guide*, 2005, Los Angeles, CA: Canon Communications, pp. 116–121.

little real understanding as to the particular type of error, its underlying cause, and its actual frequency. The very purpose of the analysis may be defeated by the arbitrary classifications of data into a few broad categories.

The frequency may be for only one system. But the *exposure* may be for many products, processes, or procedures. An exposure estimate may be in terms of the overall number of people exposed to a particular hazard or risk. For human error, it may be the frequency based on all people who would come into contact with the objects or interfaces that are inducing or permitting error.

Severity estimates

The severity of the consequences is often classified and assigned to one of several qualitative categories, for example, negligible, marginal, critical, or catastrophic (MIL-STD-882). Severity might be classified as acceptable without review, acceptable with review by management, undesirable and requiring approval by designated management authority, or unacceptable (based on a combination of frequency of occurrence and severity of the consequences).

Severity may be in terms of personal injury (no effect or negligible harm, minor injury, severe or major injury, and death). It may be in terms of property damage (little value, some value, major value, and high value).

A *detectability* range may be used, such as obvious, noticeable, obscure, and undetectable. This may be used to combine frequency, detectability, and severity into a risk estimate. Detectability refers to the discoverability of latent or hidden hazards that could become manifest unexpectedly. A *correctable human error* is one that is routinely detectable, is easily correctable by the exposed person, and may or may not have an ascertainable residual risk.

The use of just a few categories for the assessment of severity reduces the value and usefulness of the overall analysis. Detail is favored. Objectivity is essential.

Analogies and reciprocity

The analytic techniques used by other disciplines may resemble and parallel the human error analyses. The similarities may be superficial, but productive

Table 3.2 Failure Matrix

Frequency or Failure Probability		Severity Level	Detectability
1	Remote (<1)[a]	No effects	Almost certain
2	Very low (7)	Slight inconvenience	Very high
3	Low (67)	Inconvenience	High
4	Moderate (500)	Major inconvenience	Moderately high
5	Moderate (2500)	Some harm	Moderate
6	Moderate (12,500)	Daily living affected	Low
7	High (50,000)	Health impact	Very low
8	High (125,000)	Serious impact	Remote
9	Very high (333,333)	Long-term disability	Very remote
10	(>500,000)	Death	Almost impossible

[a] Incidents per million opportunities.

Source: Grossly adapted from Reid, R.D., *Quality Progress*, April 2005, pp. 90–93.

analogies often can be made. The apparent likeness may offer a bridge for cooperation and reciprocity that could be of mutual benefit. An example is the failure matrix (Table 3.2) used by quality and reliability engineers.

It seems obvious that the human error specialist should work cooperatively with allied specialists in quality, reliability, and safety. There may be differences in analytic techniques, the content evaluated, the classifications, and the ultimate objectives. However, there are similarities that could and should foster a symbiotic relationship, provide some shared content, and result in time and cost savings. Human errors that result in personal injury are also a safety problem. Human error countermeasures may involve design improvements in reliability, maintainability, and availability. The *error chain* may be a very important concept for one discipline and *situation awareness* for another. A *failure to confirm* machine responses, after control input, may be an important cause for one discipline and *mean time to failure* an important cause for another discipline. Some sharing and enrichment among disciplines is far better than orgaizational isolation and barriers to communication.

An error chain (sequence of human errors) may be associated with *tight coupling* (too little allowance for error, usually a common error). Tight coupling, it has been claimed, occurred in the Bhopal chemical plant disaster, when four safety systems failed sequentially (Bruner, 2005). The common fault was disrepair, an organizational error, suggesting that one discipline (design) did not properly communicate with another (maintainability) to ensure that one fault (disrepair) could not proceed unimpeded through a chain of events to disaster. Adequate communication between disciplines, each having unique knowledge, is usually not achieved by just a single design review session that includes all relevant disciplines.

Communication is based on long-term collaboration.

Risk

In terms of the selected risk criterion, what is important is the predicted risk under status quo conditions (the benchmark), which is compared to the amount of risk reduction (*modified risk*) achieved by various remedies, the final residual risk (*manifestable risk*), and that risk considered acceptable (*tolerable risk*). Analysts may weight (give numbers to) frequency estimates, severity estimates, and combinations thereof for risk decision-making purposes. There may be differences in terminology, such as the use of the words *mishap* and *danger index*, or the specified language of a design manual, applicable contract, specification, standard, instruction, or a directive. It is the procedure used, not the words, that is important.

A common risk classification is minor (acceptable), moderate (action required), and major (intolerable). A commonly used objective for risk mitigation is ALARP (as low as reasonably possible).

One form of risk assessment, the popular *cost–benefit analysis*, has commonly suffered from escalated benefit estimations and deflated cost estimates. This may be because the proponents of a benefit may have been forced to justify an expense by performing a cost–benefit analysis. Low cost and high benefits could be an honest belief, as influenced by an idealized halo effect. The use of fudge factors or contingency adjustments may or may not be intentional. Credibility of the source may overcome a suspect type of risk assessment, particularly if the figures seem realistic, are detailed, have a historic basis, and can be checked in some way. It is unfortunate, but the fact remains that cost–benefit conclusions are often as optimistic or pessimistic as those who have them prepared.

If the risk assessment includes an evaluation of *corporate liability*, particularly corporate manslaughter in some countries (Forlin and Appleby, 2004), a fear factor could predominate. The exposure to such legal liability should be assessed as modified by actual local government enforcement, penalties, and viable defenses. It is a subject of considerable interest to corporate management, but may be delegated to the corporate legal department whether or not it has proficiency in this area.

The *risk utility* concept balances the utility of the product, process, or service against the gravity of the risk or danger. The risk is acceptable if the utility outweighs the risk. It generally involves risk balanced against *social utility* and is performed in a manner similar to risk–benefit and cost–benefit evaluations.

The *reasonably practical* concept pertains to what degree of control or what extent of countermeasures is both reasonable and practical under the circumstances. This is somewhat similar to the ALARP risk reduction objective. It is somewhat comparable to the excessive preventable risk criterion for unacceptable risk or the simple categorization as to a defect or danger in some legal jurisdictions.

The *risk–action* approach serves to tailor the preventive or corrective action to the estimated risk levels. The indicated actions might be an immediate action priority, a high-priority action, a moderate-priority action, a low-action priority, or a no action required.

The *risk management* approach deals with estimating risks as a means of prioritizing or allocating available resources to reduce unacceptable risks.

The *managed risk* concept assumes that there will be effective procedures or actions exerted to control the risk. It is a safe use or user responsibility orientation.

The *unavoidable risk* is that which cannot be further reduced and is outweighed by political, social, or financial interests.

The *precautionary principle* in risk assessment was adopted by the European Commission. Its premise is that a danger may exist whenever there is *reasonable grounds for concern* over potentially dangerous effects on humans, the environment, animals, and plants. Thus, the European Union concluded that a danger can exist despite insufficient, inconclusive, or uncertain scientific evidence. In contrast, the U.S. Supreme Court has decided that there must be reliable scientific evidence to prove a harm. Harmonization, because of world trade needs, may result in a compromise or new criterion in the near future.

An *expected loss* may be derived from quantitative or qualitative risk estimates. A project manager may be interested in only the total losses that can be calculated, so that either preventive action can be taken or design approval granted. The loss estimate is calculated by utilizing the error frequency and incident severity to determine the consequences, which are then expressed in total dollars, injuries, or mishaps for a given set of operations, flight hours, or defined service life. It assumes some tolerable or acceptable level of loss. If everything is reduced to dollars, it may mean placing a monetary value on a human life, a procedure often criticized as to the amount or the moral implications. Fixed costs or limits on property damage or business interruption are less objectionable. For example, for a 1-year operation, the frequency of error may be classified as probable (more than 10 and less than 100), and the severity may be classified as critical (permanent partial disability or an equipment loss of $250,000 to $1 million). Ten or more losses up to $1 million each may be acceptable in military operations, but unacceptable for a civilian industrial assembly line. An expected loss estimate is something that should catch the attention of top management, be given a review by the legal and risk management department, be seriously considered by responsible procurement officials, and be evaluated by appropriate marketing directors. The loss is only an estimate that may vary appreciably in accuracy and may be subject to significant reduction with the refinement of a product or system during subsequent design improvement, development, and testing. However, it is used because it is direct and simple to understand.

Other methods of risk assessment may be found in government standards (such as EN 1050), trade standards (such as ANSI B11.TR3-2000), and various implementing regulations under laws pertaining to a particular product, process, or workplace. Such standards, regulations, or guides may use the term *risk analysis*. This is a very general term encompassing many different procedures; for example, the purpose may be simply to allocate available resources proportionate to the degree of risk involved. This contrasts with the term *risk assessment*, which denotes a characterization of risk, for example, that risk can be classified as minor, requiring no action.

Worker (OHS) risk assessment

Risk assessment is central to the occupational health and safety (OHS) standards that are used internationally. Those standards have agreed-upon basic principles of integrated system management that are considered to be the same for safety, health, quality, and the environment. Integration of those company functions has been advocated and encouraged by the publication of various international standards. For example, an international consortium, of groups from various countries, provided interim guidance for occupational or workplace health and safety assessment. This resulted in the British standards BSI-OHSAS 18001 (in 1999) and 18002 (in 2002). In these standards, risk assessment is an overall process of estimating risk and determining whether it is a tolerable risk. The International Labor Organization has indicated that there are no significant differences between the British standards (18001 and 18002) and their guidelines (ILO-OSH, 2001), except for the ambiguity in eliminating the hazards "where practicable" in the British standards.

ILO-OSH, 2001, Section 3.10.2.2, states that "a workplace hazard identification and risk assessment should be carried out" before any internal change and done "in consultation with and involving workers and their representatives." Section 3.10.1.1. states that "order of priority" in terms of risk control is (1) eliminate the hazard and risk, (2) use engineering controls or organizational measures, (3) design safe work systems, including administrative controls, and (4) provide personal protective equipment for residual risks.

In one study of root cause analysis, applied to hospital workplaces in the U.K., risk assessment was the dominant risk control failure. More than 48% of the control failures were in risk assessment, including implementation, measuring, reviewing, planning, and communicating (OSHAS, 2004). The methodology dealt with "management root causes associated with risk control system failures." A follow-up study found the same management problems. There was an inference that the main cause of incidents was that risk assessments "had not been implemented sufficiently robustly."

In essence, the organization of workplace or occupational safety and health should conform to the British standards (BSI-OHSAS 18001, 1999;

BSI-OHSAS 18002, 2000), which are more comprehensive than the OSHA (U.S.) Voluntary Protection Program (VPP, 1982) and related requirements. Compliance with the British management systems standards, including registration, and the ISO 9001 quality systems standards is compatible with the risk assessment and control procedures described in this book. Compliance, by safety and quality specialists, may be considered an elementary but necessary step toward an effective and compatible human error reduction plan.

The frequency and magnitude of occupational fatalities, injuries, and damages in all countries are grossly unacceptable. That is why there has been a historical succession of initiatives dealing with OHS (occupational health and safety) issues by the constant publication of safety and health codes, standards, and regulations; the continuing passage of legislative laws and broader common law judicial interpretations; and the creation of various government agencies empowered to inspect and levy monetary and criminal penalties. This proves that OSH is not a simple matter. There should be greater recognition that OSH requires interdisciplinary intervention, advanced knowledge, and professional skills. A far more sophisticated approach could be productive, such as that outlined for human error in the contents of this book.

One important consideration in using internationally accepted system management standards (assurance documents) is that they provide common practices (state of the art) and general acceptance in an age of multinational enterprises, world trade, and notions of social responsibility. As part of such efforts, there may be a more incisive and deeper probing, such as by root cause analysis and rigorous task analysis, that can identify unsuspected human error problems. In other words, these activities are a good data source for the human error specialist. In addition, such problems often need the special tools, techniques, and knowledge of the error specialist if they are to be appropriately resolved.

Assessment of control measures

After the implementation of an error control measure, there should be a careful monitoring or a test of the hazard in order to assess the effectiveness or ineffectiveness of the control measures. It could be a postproduction evaluation to determine the remaining (residual) risk and the need for possible control upgrades. This effort should be described in a written risk management plan so that it will not be overlooked by new or different analysts or those in a different department where responsibility has been transferred, such as the engineering department during early design and the manufacturing, quality, or sales department at a later date.

Having an overall or company risk assessment plan and file is good practice, a recommendation in some trade standards (ISO 14971), and a legal requirement in some countries. The risk management plan is an active updated document in the risk management file. The file should contain a list of government regulations that are applicable and how compliance was achieved, the trade standards that are relevant and how conformance was determined, the verification testing accomplished and the results, the certification marks self-determined or given by nationally recognized independent laboratories, and any deviations, exceptions, or marginal performance indicators. Human error reduction may be part of the objectives of the plan or part of the testing that is included in the file. All of this is valuable information for the human error specialist. For example, in the plan or file there may be a scheduled in-field or at-client operational test of an entire system, including other products and support equipment, real user personnel, and the actual environment with its stresses, demands, and conditions of so-called normal use. An assessment of control measures simply answers the question: Does it work in real life, given the human behavioral diversity that is foreseeable and predictable?

Whistle-blowing

When a supervisor or manager demands that a specific peril to others be ignored or reclassified to a harmless rating, the ethical consequences of compliance should be considered by the subordinate. There may be a legal Nuremberg obligation (see Chapter 12) to forestall a known threat to the general public given the opportunity and means. Those with knowledge must think for themselves and take appropriate action despite chain-of-command orders (that is, just following orders). However, if the supervisor's logic is good, seems defensible, and utilizes reasonable judgment, the order may entail greater benefits than risks. To act as a whistle-blower, the threat of harm should be real and substantial, since the history of whistle-blowing indicates that retaliation does occur and that the old proverb "You can't fight city hall" still has weight. Thus, the essence of the Nuremberg obligations, which require the protection of the health and safety of innocent third parties, should be achieved by means other than direct confrontation with management. It should be noted that there is an inherent conflict between the undivided loyalty due the evaluator's employer (company) and the practical social necessity of presumed loyalty to the work supervisor or program director.

Managerial actions

An aerospace reliability manager performed government-funded studies on a major subsystem during both its late design and early development stages. He limited the reliability analysis to the subsystem itself and did no system interface, safety, or human error analysis. The manager reported only a few nuisance problems, and this was welcomed by other design engineers and the project director. He apparently had decided on a short-term optimism that might facilitate his transfer to another, better project. Later, his omissions resulted in costly avoidable problems. He had committed an intentional *managerial human error* motivated by personal goals.

The manager has an ethical obligation to supervise in a manner that benefits those ultimately requesting the work and who are paying the bills. This principle may be involuntarily suppressed in the concentrated effort to meet schedule deadlines, achieve program objectives, and come in under budget. But the ethical obligation remains and retroactive remedies should be considered.

Political risk assessment

Risk assessments may have political overtones and complications when performed within a company bureaucracy. If there is governmental oversight or potential regulatory involvement, there may be compromises to accommodate the opinions of those with varied vested interests or agendas, demands for further substantiating information, unexpected complications, conditional acceptance, limitations, delays, reviews, and possible rejection. In a political matrix, anything can happen and does.

Automobile drivers may commit human errors of omission or commission that result in vehicle accidents when inattentive or distracted from the driving task. Political action has been taken on what appears to be the main culprits. Legislators around the world have enacted bans on the use of handheld cellphones while driving. However, hands-free technology (earpieces, headsets, or instrument panel devices) still permit the drivers to become so engrossed in dialing and conversation that they become inattentive to other events happening around them. Multitasking is apparently more difficult for teenage drivers than older motorists. Hands-free does allow two-hand driving, which is a benefit compared to one-hand driving with handheld cellphones. Another distraction is a driver's video monitor for television or navigation guidance displays, which may be designed to be inoperative while the vehicle is in motion, but can be easily rewired in aftermarket shops to play while driving. The executive branch of various government entities has attempted, by means of police citations and penalties, to discourage driving while eating, reading, applying cosmetics, combing hair, drinking, reaching into the rear of the vehicle, adjusting controls with one's head down, and engaging in other risk-taking behaviors that create inattention and distraction errors, such as driving with modification to the steering wheel.

There is no doubt that risk assessment does have a place in political activities, both within a company and where social (human-to-human) hazards create external political risks. It is not that difficult to extend risk assessment to more purely political matters because generic techniques have wide applicability. Generically, the risk assessment analytic techniques are broadly comprehensive, very detailed, mostly objective, risk based, and problem solving, with a focus on hazard avoidance alternatives, and are system interactive oriented.

Postmarket considerations

Risk analyses should include references to the applicable regulations, such as the European Union's proposed REACH directive (the registration, evaluation, authorization, and restrictions of chemicals). This 1700-page draft regulation directs that each of some 30,000 chemicals should be tested and then classified as either authorized or restricted (dangerous). Most chemicals used in the U.S. and in other countries have not been appropriately tested as to toxicity and have not had adequate risk impact studies performed. A risk analysis should predict probable future regulatory requirements, including any retroactive effects that might necessitate product withdrawal from the market, possible recall, or any remanufacture or update requirements for noncompliant items. For example, one commonly used herbicide designed to kill plants (a weed killer) was considered nontoxic to animals and generally safe to use. Years after the herbicide was first marketed, it was found to be lethal to some amphibians, such as frogs. Spraying the herbicide on ponds and shallow bodies of water then became an unauthorized use, a misapplication, a misuse, and an error. The active ingredient of herbicide in the chemical formulation was not the culprit — it was the surfactant that permitted the chemical to penetrate leaves.

This herbicide problem illustrates the need for postmarket surveillance in an age of increasingly rapid communications between consumers, manufacturers, government agencies, and the public media on a worldwide basis. Concepts involving human error are central to many postmarketing problems, so a risk analysis dealing wth human error should be constantly updated to reflect adverse events, identified human error problems, corrective action undertakings, and efforts at continuing product improvement. The discussions of relevant global harmonization task forces may provide access to a more consistent set of requirements for safety, quality, and performance. But there are constant attempts to consolidate and standardize, attain consistent and common technical specifications, improve quality system requirements, achieve good manufacturing practices, and lobby for less burdensome notification and submissions to a wide variety of government agencies. If postmarket human error control is important, there should be access to the background information needed for an adequate risk analysis involving human error.

Note: The term *mishap* (an unplanned event or error) may be used to denote the initiator or trigger that unleashes the unwanted effects or consequences of a hazard. The hazard has the potential for causing injury or damage to a target of opportunity (a person or object proximately exposed). The trilogy of *hazard* (potential harm), *risk* (the severity of consequences), and *danger* (unacceptable risk) remains embedded in several disciplines regardless of the terminology utilized.

The term *human risk engineering* may be used to describe activities related to human error control.

Caveats

Conflicts

In the process of a risk evaluation, should some *conflict of interest* or adverse relationship arise, this potential bias should be disclosed to the immediate authority responsible for the tasks involved. This includes issues relating to prior employment, stock ownership, familial relationships, near-future prospective employment, and an early completion bonus or stock option.

Objectivity

Preference should be given to *objective data*, a lesser reliance given to credible subjective opinion, and minimal faith given to pure guesswork and quick speculation. That which involves a serious lack of proper foundation, a lack of informed extrapolation, or uneducated personal opinions should be disclosed so that others who might rely on such estimations are properly cautioned.

Secrecy

The risk evaluation actions and outcomes should be treated as confidential, *proprietary,* and restricted dissemination matters. An attempt should be made to preserve what could be considered trade secret, until and unless good reason and proper authority provide for its release, declassification, modification, or destruction.

Documentation

Appropriate *documentation* of the risk assessment process should be initiated, maintained, and supplemented as necessary with an accountability form of dated sign-offs. The records should have sufficient clarity for general understanding and be appropriate for others to review, extend, and modify. The records should be held, in some safe and accessible location, for a stated period of time beyond the duration of the service life (as modified) of the affected products, processes, systems, or services. Such records may have value and utility in future risk assessments, lessons-learned design analyses, liability defense actions, and in formulating future corporate management plans and procedures.

No analysis possible

There will be situations in which a human error problem has already been discovered and there is a request or demand for an immediate recommendation as to a possible solution. Time may be of the essence, so the request may not be unreasonable under the circumstances, and no time-consuming analysis is possible. For an improvised, *ad hoc,* or interim response, the error specialist could go directly to the middle of this book and use Chapter 6 as a checklist of possible remedies. This could be supplemented with the guidelines, ethics, and managerial chapters. This assumes that the lean temporary error-corrective measures will be subsequently enriched by further studies and follow-up tests or audits to determine if the provisional remedies are effective.

chapter four

Alternative analytic methods

Introduction

There are many forms of risk assessment, ranging from the worthless to the extremely valuable. What may be a productive approach in one set of circumstances may be a blunder or a charade in another situation. Selection of the method most appropriate is the key initial choice in developing a suitable risk assessment policy and plan based on need, cost, technical difficulty, available competencies, the feasibility of instituting preventive and corrective remedial actions, and due consideration of professional ethics. The following alternative methods may suggest the elements that can be adapted for a tailored or custom-fitted written risk assessment plan.

No data methods

There are many situations in which there are no relevant data, a lack of credible data, serious conflict about the available data, or just no opportunity to gather and assess data because of time, cost, or urgency considerations. There are several no-data methods used to formulate provisional, temporary, or interim risk assessments.

Default decisions

If no data exist, the specialist could employ the most conservative approach by using a *worst-case assumption* or fallback position. This protects the operator, user, consumer, and company or agency in terms of public health, prospective liability, and market reputation. As field experience accumulates, experimental data are obtained, or test verification occurs, a more realistic estimate can then be made.

The real-life problem with this approach is its drastic conclusionary nature, which may stall the development process or the marketing of a product. Its benefit is to force a meaningful proactive analysis that might otherwise be overlooked or forgotten. This drastic default assumption recognizes the temptation to give a passing score in terms of risk, rather than delay product development. That score may or may not be corrected in a timely fashion at a later date. It may not be marked as an interim or preliminary estimate that needs verification or peer review. In essence, the actual risk is uncertain; it could be too high or very low. A passing score or acceptable risk designation may reflect a passive rather than active analytic approach. It could imply that further action can be overlooked. It just contains an unacceptable degree of uncertainty. It serves to bypass rigorous evaluation, and the passing score can be misinterpreted as a true worst-case scenario. The passing score approach perverts the default assumption method and should be exercised with great caution and only where necessity dictates.

The precautionary principle

A less drastic approach to risk assessment is to use the precautionary principle, which suggests that *appropriate caution* should be exercised when there is any significant uncertainty about a risk. In other words, use caution whenever there are reasonable concerns about potentially dangerous effects that could result from human error. The precautionary principle is applicable in situations where scientific evidence is insufficient, inconclusive, or uncertain.

When the European Commission adopted the precautionary principle, it indicated as policy in risk assessments that a danger may exist whenever there are reasonable grounds for concern over potentially dangerous effects on humans, the environment, animals, and plants. What has been disputed is the exact meaning of "reasonable grounds" and "dangerous effects." Is danger anything more than, or some multiple of, a *de minimis* (insignificant) risk, an acceptable risk (less than a one-in-a-million chance), or a relative risk (an excess over the background risk)? If categorized as serious (capable of widespread harm), is the precautionary principle sufficient to suggest a high-priority *action* and resolution of an unacceptable level of risk? Application of the precautionary principle is often necessary to counterbalance or offset the actions of those who manifest a high tolerance of risk.

In contrast, the U.S. Supreme Court has decided that there must be reliable scientific evidence to prove a harm. This is a very high and difficult level of proof. One explanation of the difference between the precautionary principle (no evidence) and the required proof of harm (reliable evidence) is that one concept is proactive (premarket) and the other reactive (postmarket). That is, the precautionary principle deals with predictable, foreseeable, or future harm, and the reliable evidence rule deals with actual harm that has occurred or claimed damage already accrued.

Expert elicitation

When there is little or no applicable data to serve as foundation for a risk assessment, an independent and diverse group of specialists might be called upon for their expert judgments. This reliance on outside experts has as its objective a drawing out or discovery of the truth about all of the risks involved. This is generally accomplished by a frank discussion and consensus judgment. The process is called expert elicitation or education.

The group may critically review the available data, past company experiences, and the personal judgments of others. It may develop risk probability estimates, dated probable outcomes, and descriptions of the means by which its opinions can be tested and verified. The judgments elicited from experts constitute only a preliminary estimate, not a final determination of risk. They can be valuable in the absence of supporting information, where there are conflicting subjective opinions, or where the primary vested interests should be counterbalanced or evaluated in some fashion to ensure reasonable accuracy in subjective risk assessments.

Local consensus

Utilizing a group of outside experts may be somewhat unpalatable to the responsible design or development engineers. It may be perceived as a personal affront or something not worthy of the time and cost. A somewhat similar, but *local consensus*, approach might be found in a company's formal design review process, the use of in-house focus groups, or the assignment of technical committees to attempt to resolve defined business uncertainties and unknowns, to address specific and difficult problem solving, to look at habit or recidivistic concerns, or other tasks. It may be more acceptable and productive to organize a diverse group of in-house specialists with an appropriate knowledge base, an absence of immediate bias, and some personal interest or motivation to help other internal organizations. One disadvantage is the uncertainty of relying upon others who may be considered strangers even if they are co-employees. An advantage of local consensus is that it may be necessary to overcome a design unit's pride, arrogance, and a refusal to acknowledge that any meaningful human error could ever occur or, if so, be mitigated in a reasonable fashion. A well-organized and carefully selected internal focus group may accomplish the same objective in a more informed manner. The ultimate objectives of a local consensus approach are to help identify and categorize human error, reduce uncertainties, and reduce predictable harm.

Error troubleshooting

The mere suspicion of a cause of human error may trigger some effort to investigate and resolve the matter. The suspicion may arise from a finding, in an accident reconstruction report, that human error could have been a contributing factor in the accident causation. Press reports of accidents frequently reflect first impressions that someone did something wrong, a person made a mistake and was at fault, or someone committed a human error. The first impressions, even if corrected or superseded, should be investigated. In fact, almost any significant trouble report requires some *error troubleshooting* because the error causation may fade away and then some time later reappear unexpectedly. The late resurrected human errors, if there is no written record or investigation, may be very difficult to explain, defend, or refute.

Troubleshooting is often an attempt to remedy a no-data situation. Just a hint of a human error problem arises, and this may initiate an attempt to get more explanatory information. The error troubleshooting may be conducted on site at an accident, by a review of records and telephone interviews, by a professional-level evaluation at a distant office, or by analytical troubleshooting using mathematical models to help determine causation of the human error. It may be conducted by one troubleshooter or a carefully selected human failure investigation team. The sequence may be to gather data, define the problem errors, organize the collected information, formulate accident cause scenarios, test cause hypotheses, identify possible solutions, evaluate remedies, and implement corrective action.

The availability of hard data on human error serves to increase the burden of proof on those promoting alternative failure causation and those exercising so-called common sense in the absence of valid or meaningful data. Error troubleshooting is a deliberate effort to remove the uncertainties in human error categorization, to enable and facilitate transfer of human error knowledge to new situations, to determine appropriate error countermeasures, and to give enhanced standing and credibility to human error specialists.

Search teams

Where there are no available data, it may be advisable to conduct informed search expeditions into operations, such as manufacturing, to see how work tasks are actually being accomplished. Very often the conditions that can induce human error are obvious, may be well known to the affected workers, and may be awaiting resolution by work supervisors who are afraid of personal unilateral changes, unaware of the possible remedies, or who need support or funding assistance. In some factories, improved procedures may require remote headquarters approval, be almost always deferred to the next fiscal year, or be habitually relegated to scheduled new model changes. Labor union representatives may become involved in any identified worker problem and could help or slow improvements for a variety of reasons. Stated another way, bureaucracies develop in most social situations and often hinder constructive change. The human error search team should be perceived as outside the control of the affected organization, overtly helpful in character and action, and nonpunitive in intent or procedure.

In some companies, it often seems as though almost everyone aspires to become a manager, act in a similar role, or become a staff associate to a manager. Engineers are usually educated and trained to act in supervisory roles, rather than in a hands-on manner. This may result in limited learning experiences and a form of isolation in one specialty, office, or organizational niche. Mistakes may be made because of a lack of understanding of what goes on in other units, activities, or divisions of a company. A design engineer may not understand the needs of a manufacturing engineer, a manufacturing engineer may not understand the needs of the marketing specialist, and the marketing specialist may not understand the actual needs of the customer, component integrator, or product assembler. A failure to understand tends to breed inconsiderate or improper decisions, including making easily avoidable human errors. For this reason, a search team should include persons who may need or could benefit from real hands-on experience in other parts of the company or industry. It is an opportunity to learn generally as well as to collect specific relevant data.

The use of a search team is also illustrative of the personal initiative and individual creativity that is often necessary in no-data situations. The human error specialist may have to devise a means to collect usable data, and this may produce unique information not available in other data sources. It is also a test of negotiating skills, to maintain good interpersonal social skills on the floor while helping devise and institute improved error-reduced work procedures.

Systems integration analysis

The prime method described in the preceding chapter involves a rather detailed and comprehensive analysis procedure. An analogy would be the use of a high-powered searchlight to illuminate each of the elements, units, or parts of a system and their connections and interfaces — that is, how the system is integrated, how it functions, and what opportunities are presented for human error. It is significant that everything is put under a virtual microscope to determine the cause, effects, and countermeasures that could be effective in reducing the frequency of error, mitigating the severity of any consequences, and uncovering or avoiding side effects.

When this methodology deals with system integration, it is best performed in early design. It is generally costly and extends over a long period. It is highly predictive and deals with reasonably foreseeable events. It may not be appropriate for all situations. There are alternative methods, but the essence of the procedure should be retained because of its incisive manner of dealing with the cause and effect of human error.

A system is considered a complex set of interacting parts and assemblies intended to perform a given function. It may be a large *macrosystem* such as a submarine and its land and sea support arrangement, together with a wide assortment of people assigned many different tasks. It could be a ballistic or space system rocket engine with its ground support equipment and the people who operate and maintain it. It might be a commercial aircraft integrated into a complex of other aircraft, with flight crews and personnel involved with ticketing, scheduling, and maintenance. But the word *macro* could also apply to items of small physical size, such as a digital camera with precise interacting parts that are mechanical, electrical, electronic, software programmed, connected to a photoprocessor or copier, and with auditory and visual displays to the picture taker. Familiarity with a system tends to make it seem macro, even when the first impression is that of an average microsystem or something less than a real system. Experience tends to magnify complexity, interactions, and the system aspects of design elements.

Successive evaluations

It may be appropriate to perform a human error analysis in several stages. A *preliminary* evaluation might be conducted when there is a shortage of data, limited needed information, or some of the major design decisions are not finalized. A preliminary or draft version can always be upgraded, extended, or made more precise in the design development and test stage of product development. This permits time for special studies and relevant testing of a multidisciplinary character that could provide answers to critical questions. The final, or even an aftermarket, version of the analysis could be performed when the production versions of the product are released, as-built drawings are available, initial market responses have occurred, in-field data have become available, significant modifications have occurred in manufacturing or product assembly, or there are customer fixes or other changes in the product. It may be that significant human errors often become manifest at a late date, so an updated successive analysis may be of great value. Latent human errors are very important and should be discovered, identified, and recognized as problems early during customer use.

Another form of successive evaluation is when there is periodic redesign. For example, each model year of an automobile or a continuing line of farm equipment may be modified periodically according to the needs of the customer marketplace. This may be called trim changes not needing extensive testing, superficial customization that does not change the basic design, or normal configuration modifications for a continual improvement program. But the changes may induce human error, and successive evaluations should be considered.

Targeted errors

There may be some rather narrow issues as to a particular known human error. The issues may not be time critical as to a project schedule. A broad-scope inquiry might not be cost justified. For example, it could be a prior abuse of an electromechanical control in one geographical location, a history of forgetful behavior in dealing with a product, or an unacceptable error rate that continues unabated despite a commonsense fix. This may require a directed, focused type of an observational technique or a traditional experimental method of uncovering the identity and weight to be given to relevant independent and dependent variables. There is often a need for immediate problem resolution when a particular identified human error is targeted by management for immediate attention for purposes of customer satisfaction.

Immediate problem resolution

Human error problems may arise from discussions with project engineers at the last moments in the design process (with deadlines), from design reviews (with schedule problems), from product integrity testing, from unexpected customer or consumer complaints, from field or service reports, or from accident investigations. It may be just one identified problem, but with an express or implied request for a quick fix.

In the rush to help solve the problem, utilizing the *substantial factor* test of proximate causation, it is easy to jump to conclusions on one factor alone, then communicate a marginally effective recommendation. It is far better to use a method of *forced detail* that provides a better understanding of all relevant contributing factors. It is the whole picture that is important, not just a quick close-up look at the tail of an elephant, which may be misleading.

For a forced understanding on one problem, the essential methodological steps described earlier in this book should be utilized. For immediate problem solving, this may be translated as the FODISE approach:

F — *Frequency* of human error; including a reference to the total number (N) of opportunities for error events, operations, or transactions, for example, 17 errors per 1000 operations.

O — *Opportunities* for harm by excursions into a hazard zone (including near misses), for example, 15 errors in the hazard zone.

D — *Detection* and correction, before harm, by the operator or the system, for example, only one remaining error causes harm.

I — *Injuries* and damage that result from error, for example, one finger amputated.

S — *Severity* of the consequence of the error; see below.

E — *Effect* of current or proposed countermeasures; see below.

The amount and vigor of the efforts at problem resolution very often depend on the social context in which the injuries occur. If an injury occurs in a workplace, generally there is no-fault workers' compensation insurance that provides a financial remedy to the injured person. Investigations, if any, are limited, conclusionary, and in the language of the applicable statute, insurance plan, and occupational standards from state and federal agencies. Only rarely are there *third-party actions* that may serve to communicate human error problems to equipment manufacturers, general contractors, the employer management, or regulatory agencies. *No fault* means silence on real causation. This suggests that human error investigations should be direct real-world observations, personal interrogation of exposed workers, and informed machine evaluations.

A somewhat similar situation exists for motor vehicle collisions, where casualty insurance covers the claims. Accident reports, if any, use stereotyped language from the applicable vehicle code and driver's handbook. Accident summaries often use very general language, have predetermined categories, and result in politically acceptable but biased conclusions. There may be exceptions, but the insurance remedy serves to mask the specifics of human error causation. Just a few in-depth, immediate, and focused (on a relevant issue and product) accident investigations and independent accident reconstructions could yield valuable information not otherwise available to interested parties.

The FODISE methodology is also applicable to traffic safety situations. This could include the evaluation of a roadway intersection where some vehicles do not stop or just slow down for the red light of a traffic control signal or stop sign. The evaluation would include the system of traffic control, which includes traffic approaches, pavement markings such as lane lines and stop limit lines, advance warning signs, speed signs, off-road visual and auditory obstructions, other traffic control devices, and hazards such as off-road visual and auditory obstructions. The human error approach goes far beyond findings of fault, police citations, and insurance claims requirements, should there be an interest in more meaningful human error causation and reduction.

The human error problem may emerge from the limitations of machine design, where time-related breakdowns require human intervention in maintenance and repair operations. There may be worker stress because of the need to return the machine to normal operation as quickly and inexpensively as possible. For example, the Piper Alpha disaster, in 1988, occurred when a pump relief valve was removed from an offshore oil rig for maintenance overhaul (Cullen, 1990). The primary human error occurred when the workers on another shift started the pump, not knowing that the pump relief valve was missing.

A similar problem may occur in electronically controlled safety systems, where there is a lack of redundancy and a single-point failure occurs. This may require some positive affirmative action by the previously protected human, but acts of omission often occur because of the unexpected situation. Redundancy means that there is some provision to ensure that there is more than one way to accomplish a given function. There is some protection against component failure, such as electronic circuits that have duplicate elements with one on standby, appropriate decision and switching devices, and protection from the secondary effects of component failure. The human error specialist should know or be alerted to the possibility of single-point failures so that adequate emergency procedures and appropriate human actions will occur without unnecessary injury or further damage.

Sometimes, one human error can lead to another. For example, health care providers in hospitals have been accused of carelessness or medical error when there are high rates of pathogenic bacterial infection among their patients. The bacteria may cause urinary tract infections, meningitis, pneumonia, boils, heart muscle infections, and episodes of flesh-eating bacteria, and such infections can be life threatening when the patient's immune system has been compromised. The bacteria include *Staphylococus aureus, Staphylococcus epidermidis, Streptococcus pneumoniae, Enterococcus faecalis*, and nearly 50 other types of bacteria. A lesion may require a differential diagnosis between an infection and an inflammatory process. But surgical biopsies carry a risk of medical error. Magnetic resonance imaging (MRI) may detect an inflammation, but other imaging with radioactive tracers can provide signals from an infected spot even within a cell. The pinpointing focus narrows and guides the biopsy, thus reducing the risk of surgical error and unnecessary biopsies. Less invasive techniques also may help prevent medical error during the treatment of other disease entities. This includes alternatives to the use of contrast agents, such as photoacoustic imaging of blood vessels to monitor anti-angiogenic cancer treatments. The blood hemoglobin absorbs infrared laser light, resulting in overdilation of red blood cells, local temperature increases, and increased local blood pressure, which produce ultrasonic waves that enable the constructs of visual three-dimensional maps of the blood vessels.

It should not be forgotten that human errors are expressed in a *macroergonomics* context, that is, a human–organization interface (Hendrick and Kleiner, 2001). This means there should be an appreciation of the larger system. In other words, the focus may be on immediate problems resolution in a specific situation, but the perspective includes the interface between the individual and other system elements and interfaces. The macro or big picture is ever present as a backdrop, even when the problem solving has a narrow focus on the micro or pinpointed problem area.

Uncertainty analysis

An early measure of success for the analyst is a measurable *reduction* in human error propensity, particularly for error associated with unwanted or undesirable consequences. Equally important is the *remaining error*, the uncontrolled or residual human error.

It is the residual propensity for error that should be of serious concern, since it may be the cause of unexpected future perturbations in the function of a product, process, procedure, or system. The question may be how could it be measured if the residual risk includes the unknown human errors yet to be discovered or to become manifest.

One category of residual risk is that which is known and accepted by decision or indecision. It is a voluntary *assumed risk* of harm. It may be an acceptable risk under a risk–benefit analysis, the calculated risk remaining after an error correction, that risk subjectively determined to be tolerable and acceptable, or a risk reduction in an unverified or unvalidated attempt at error reduction. Whatever the source, residual risk is an indicator of uncertainty.

The undiscovered residual risk obviously cannot be measured. It might be subject to guesses based on the kind of analyses that have been performed. This uncertainty sometimes can be cured by a life cycle monitoring effort, that is, a continued error discovery and remediation program.

The outcome of an uncertainty analysis reflects just how good a job was done by the usability and human error analysts. More important, it is a legitimate concern of management, the customer, and the user since uncertainty is equivalent to gambling. Good management wants to reduce uncertainties that might cause subsequent problems. The human error analysis is fundamentally a way in which uncertainty is minimized to acceptable levels in the type, frequency, severity, and adverse outcome.

Exclusions

There are circumstances when entirely different methods may be advisable for human error prevention. Such methods may have limited data acquisition needs, a narrowed risk assessment, and a focus on a restricted subpopulation. If the objectives are different, the approach to prevention may be different.

It may be an important company objective to select and locate managers who will not commit errors in the implementation of company policy and in the application of defined business practices. There may be concern about harmonious interactions between managers, aggressive achievement of performance goals, proper negotiating strategies, due regard for company values, and the maintenance of a desired company cultural climate. For further information in this book, see Chapter 9.

It may be determined that a simple screening process is necessary to weed out the most likely to commit human errors. The preferred method is structured interviews, tailored to carefully defined objectives and conducted by qualified specialists. Written tests or questionnaires are less expensive, less effective, and cannot be quickly adjusted to follow leads such as body language, facial expressions, language cues, and undesirable reactions to scenarios depicting current human error problems in the company. The screening is to detect the probability of human errors in a specific job, the feasibility of corrective action for the individual assigned to perform certain tasks, and the advisability of an assignment to a job that is less likely to induce undesirable human error. The human–human interface is an important variable since one person can provoke errors by another person.

Note: It is important to fully understand the strengths and weaknesses of the allied disciplines with which coordination is necessary, including system safety, reliability engineering, quality assurance, and human factors. For good instructive information on current system safety theory and practices, read Ericson (2005). For information on quality and reliability, read Raheja (1991). For human factors, read Karwowski and Marras (2005).

Caveats

Just ask

One facet of the search for sources of human errors is the simple process of asking those responsible for the design of a product, its integrity, its customer satisfaction, and others having management influence. Errors important to those in the design process should become known and considered important by the human error specialist.

Subjective risk

The responses to questions may suggest some risk level, relative importance, or prioritization. It is important to remember that the degree of risk may be in the eye of the beholder, but this should not be discounted arbitrarily.

Spin

Protective rationalizations and spin may occur in response to questions about human error. Frank admissions are helpful, but so are dismissive conclusions. Behind the spin may be uncorrected problems or insufficient remedies.

Legal risk

Routinely check for information that describes various legal concepts relating to human error. For example, jurors may be instructed to decide fault or liability on the basis of whether there is an excessive preventable danger (a risk–benefit criterion). This means that what might appear, in design evaluations, to be a comparatively low risk actually may be sufficient to warrant design action, particularly if the remedy is of low cost and technologically simple. Similarly, jurors may be instructed that the product is defective if it fails to meet ordinary consumer expectations or that there is a breach of an implied warranty of fitness for a particular use. Stated another way, therapeutic intervention may be based on an alternative definition of actionable risk, a risk that may be defined by others.

Allocating resources

Estimating and characterizing risk is not an end in itself as many believe, but it is a means to rationally allocate resources to achieve acceptable levels of risk and to avoid harm.

chapter five

Behavioral vectors

Introduction

People may appear, to a design engineer, to have a certain general communality in terms of basic traits, capabilities, aspirations, and social behavior. The so-called average person may vary somewhat in appearance and have some distinguishing features, but such persons are all basically human, can learn, adapt, adjust, understand instructions, follow the prescribed rules of the road or occupation, and perform work procedures as expected. The basic adaptability of humans may suggest that the differences can be ignored or marginalized, since they can adapt, to some degree, to almost any product, process, or system. Perhaps this overestimation of human capability and lack of meaningful consideration of individual differences is a prime cause of undesired human error.

In fact, each person who comes into contact with an engineered product is quite different, and some could manifest behavioral problems or disorders that are harmful to expected human and machine performance. If the behavioral disorders are moderate or severe in the anticipated worker or user pool, there should be some countermeasures instituted as a preventive or corrective action.

This chapter may assist in the differential diagnosis of various behavioral vectors as they relate to human error.

A conduct disorder may involve violations of social norms or rules, with aggressive threats and destruction of property. An antisocial personality disorder may result in irresponsible work behavior. An oppositional defiant disorder may be manifest as a recurrent pattern of hostile, disobedient, and negative behavior toward authority figures. Such disorders usually begin in childhood or early adolescence, so they are known to others. They can be diagnosed by behavioral specialists and may be compensated by design.

Human error may be induced by physical disability conditions (with behavioral disturbances), substance abuse or intoxication, or recognized disorders of communication, learning, reading, motor skills, mood, attention, or adjustment to stressors. In terms of general intellectual functioning, by definition, 50% of the general population score and are classified as below average.

Thus, there should be no question as to the importance of thoughtfully *considering* who will actually represent the foreseeable, predictable, and anticipated user and worker population. The next step is to *expect* human error of some type, frequency, and severity of consequence. Only then can appropriate *countermeasures* be determined.

General approach

It is important to clearly identify, appropriately understand, and determine the true cause of error behavior among all people who could have the opportunity to commit significant human errors in a system, process, product, or service. That is, find out *who* could commit the error and *why*. Determining the cause may suggest whether or not a selected remedy would be appropriate and effective.

There is a broad range of behavior that is considered normal and predictable. Design safety should consider and accommodate these foreseeable human characteristics. But anticipated behavior beyond the normal may be a cause of concern and require special design attention. It may require carefully defined access restrictions, effective use limitations, or special worker selection and placement procedures.

Those who could commit relevant human error may be from a small, highly select group, functioning under close supervision, or in compliance with strong cultural norms. But they could be from the general population with little behavioral restrictions.

The following descriptive behavioral information should be used as a checklist or reminder of the behavioral vectors that should be considered in the control of human error. The analyst should serve to put it all together and make modifications as necessary to his risk formulations and countermeasures. In such activities, the analyst integrates a broad range of information, making differential diagnoses from the strength, direction, and character of the various predictable behaviors, such as those enumerated below.

Basic human characteristics

Humans consistently manifest individual differences (attributes) within a group (population). Each human also varies (variability) in the overt expression of each attribute. These differences may be masked, submerged, or partially hidden by the similarities (communality) of humans in general. In terms of human error *prediction*, we are concerned with the statistical mean (average), the variation (variance), what is considered normal by some purposeful definition, and the tolerable limits (acceptable range) of the relevant human conduct, performance, or error. In terms of human error *control*, we are particularly concerned with variance control (consistency). These considerations should influence the way in which the following human characteristics are perceived and acted upon.

Learning ability

This attribute is important in terms of efficiently adapting to initial machine operation, implementing modifications in work procedures, understanding instructions, and recognizing hazards. It may be reflected in coping skills, creativity, and error avoidance. Human learning ability may be related to success in education, training, and a quickness in adaptations to changes in social, work, and environmental situations. Learning ability may be significantly affected by other factors described in this chapter.

Communication skills

This attribute deals with interpersonal information transfer. It is generally dependent on reading skills (knowledge acquisition), verbal language ability (expression of ideas), and a willingness to exchange information with others. It may be seriously compromised by other factors, both personal and social. Typical communication disorders include articulation (incorrect sounds), voice (pitch, loudness, or complete loss), and fluency (normal flow or rhythm). Stuttering effects about 1% of the general population.

Information processing

Humans can and do process vast quantities of data, with brain mechanisms that have evolved to prevent overload (filter the clutter), select that which needs attention (suppression or focus), and help with executive functions (priming and coding for decision making). These brain mechanisms should ensure injury avoidance (self-protection) and the accuracy of interpretations, facilitate error avoidance, and minimize undesirable distractions, conflicts, and confusions. The desirable cognitive functions may be enhanced by some error control techniques or adversely affected by other factors described in this chapter.

Physical attributes

Humans vary in their physical attributes, such as height, weight, mobility, vision, hearing, smell, and touch. In addition to normal or intrinsic variation, there may be injuries that are incapacitating in some manner. A disabled person, such as a paraplegic, may retain superior cognitive skills.

Reaction

Human reaction times may range from 0.5 seconds to more than 3.0 seconds, depending on the task, attention, and situation awareness. There may be standard reaction times, such as those used in vehicle traffic accident reconstruction or in roadway design. Humans may characteristically have a panic reaction in some emergency situations or an avoidance reaction to noxious stimuli.

Language skills

The ability to understand and express words, as mediated by the left temporal lobe and part of the front lobe, varies greatly among humans. For example, many people suffer from dyslexia, a neurological learning disability, characterized by poor word recognition, spelling, and decoding ability. For example, the word *was* may be read as *saw* and *on* as *no*. Poor reading comprehension may reduce the amount of reading and impede the acquisition of background knowledge. Those who are dyslexic are usually well aware of their many human errors in word recognition.

Illustrative personality traits

This section deals with personality *traits* that are fairly common among the general population. The behavioral manifestations may be slight to moderate, but may be significant in terms of human error causation. The more severe personality *disorders* are detailed in Chapter 9 and are presented from a slightly different perspective and for a different purpose.

Antisocial personality

This conduct disorder is manifested in a failure to accept social norms, accept responsibility, provide consistent work behavior, and a tendency to violate the rights of others. Additional human error concerns are constant violations of rules, extreme misbehavior, delinquency, and persistent lying.

Histrionic personality

This consists of behavior that is intensely reactive, dramatic, inconsiderate, egocentric, and with exaggerated expression of emotions. Of human error significance are angry outbursts, irrational behavior, vain dependency, and suicidal attempts.

Avoidant personality

This is manifested by social withdrawal, social inhibition, low self-esteem, unwillingness to enter into relationships, hypersensitivity to negative evaluations and ridicule, and subjective feelings of distress and inadequacy. Human error concerns are the problems in occupational and social functioning, including avoidance of work and group relationships, preoccupation about criticism, a reluctance to take personal risks, and the avoidance of new activities.

Passive-aggressive personality

This is an oppositional disorder where there is resistance to demands for better performance, intentional inefficiencies, and a lack of assertive behavior in some circumstances. Human error concerns include procrastination, stubbornness, and the appearance of forgetfulness.

Dependent personality

This personality type is characterized by dependent and submissive behavior, a perception of inadequate functioning without the help of others, difficulty initiating activities, and tolerance of abuse. Human error considerations include difficulty in making decisions, general passivity, avoidance of responsibility, difficulty in expressing disagreement with others, and agreement with others even when wrong.

Obsessive-compulsive personality

This is characterized by a preoccupation with order, perfectionism, attention to detail, a need for personal and interpersonal control, an excessive conscientiousness, stubbornness, rigidity, and unusually high self-criticism. Human error concerns include repeatedly checking for mistakes, delays due to attention to detail, inability to follow procedures from memory when written instructions are misplaced, important tasks are left to the last minute, a reluctance to delegate tasks, avoidance of alternatives, and difficulty accepting the viewpoints of others.

Narcissistic personality

This personality type encompasses a pattern of grandiose feelings of self-importance and superiority, a need for admiration and recognition, intolerance of others, a lack of empathy and appreciation of the feelings of others, emotional coldness, and a lack of reciprocal interest in others. Human error concerns include an overestimation of abilities, inflation of accomplishments, impatience with others, harsh devaluation of the accomplishments of others, and unwillingness to take risks.

Symptom cross-reference

Achievement devaluation — See avoidant personality behavior.
Aliases — See antisocial behavior.
Angry outbursts — See histrionic behavior.
Arrogance — See narcissistic personality behavior.
Avoidance of new activity — See avoidant personality.
Best of everything — See narcissistic personality.
Boastful — See narcissistic personality.
Conning others — See antisocial.
Conscientious — See obsessive-compulsive personality.
Control — See obsessive-compulsive personality.
Debt defaults — See antisocial.
Decision-making difficulty — See dependent personality.
Delays — See obsessive-compulsive personality.
Delegation reluctance — See obsessive-compulsive personality.
Dependency — See dependent personality.

Detail preoccupation — See obsessive-compulsive personality.
Disagreement in error — See dependent personality.
Emotionally shallow — See histrionic personality.
Empathy lacking — See narcissistic personality.
Entitlement expectations — See narcissistic personality.
Enviousness — See narcissistic personality.
Exaggerated responses — See histrionic.
Exploitive — See narcissistic personality.
Forgetfulness — See passive-aggressive personality.
Gifted status — See narcissistic personality.
Hypersensitivity to rejection — See avoidant personality.
Impressionistic speech — See histrionic personality.
Impulsiveness — See histrionic.
Impulsivity — See antisocial.
Inadequacy feelings — See avoidant personality.
Inefficiency — See passive-aggressive personality.
Inflexible — See obsessive-compulsive personality.
Initiation difficulty — See dependent personality.
Irresponsibility — See antisocial.
Isolated — See avoidant personality.
Lonely — See avoidant personality.
Lying — See antisocial personality.
Misbehavior — See antisocial personality.
Mistake checking — See obsessive-compulsive personality.
New activity avoidance — See avoidant personality.
Overreaction — See histrionic personality.
Overworking others — See narcissistic personality.
Passivity — See dependent personality.
Perfectionism — See obsessive-compulsive personality.
Power preoccupation — See narcissistic personality.
Pretentious — See narcissistic personality.
Procrastination — See passive-aggressive personality.
Repetition — See obsessive-compulsive personality.
Responsibility — See dependent personality.
Rigidity — See obsessive-compulsive personality.
Risk avoidance — See avoidant personality and narcissistic personality.
Rule infractions — See antisocal personality.
Self-critical — See obsessive-compulsive personality.
Self-esteem enhanced — See narcissistic personality.
Self-esteem low — See avoidant personality.
Self-importance — See narcissistic personality.
Sensitivity lacking — See narcissistic personality.
Shy — See avoidant personality.
Social phobias — See avoidant personality.
Stubbornness — See passive-aggressive personality and obsessive-compulsive personality.

Submissive behavior — See dependent personality.
Suggestibility — See histrionic personality.
Superiority feelings — See narcissistic personality.
Timid — See avoidant personality.
Underestimation of others — See narcissistic personality.
Viewpoint — See obsessive-compulsive personality.
Withdrawal — See avoidant personality.

Normalcy and impairment

A psychometric perspective

An indication that there is average performance may suggest that a psycho-metric test score approximates the arithmetic mean, mode, or median. *Average* may convey the wrong meaning that a test score is near the center of all of the population scores. Thus, the word *normal* is used to include a much broader inclusive population.

A test administrator may use a two-category (normal and impaired) classification procedure to provide contextual meaning to others. For example, a particular raw score on a test may be transformed into a percentile score. The test administrator may then say that the test score was in the lowest 1% of all those taking the test, or that 99% did better than the subject. The reasonable inference is that 1% of the population is impaired and 99% are normal.

Another cutoff between normal and impaired is to use the two standard deviations below the mean criterion, based on a one-tail normal distribution. That is, 2.28% of the population would score in the impaired range and 97.72% in the normal range.

There are classification schemes that have seven categories, including very superior, superior, high average, average, low average, borderline, and impaired. The impaired score may be equal to or less than 2% of the population.

A relatively new classification procedure uses (negative) one standard deviation as *mildly impaired* (that is, 84.13% are normal and 15.87% are mildly impaired), two standard deviations as *moderately impaired* (2.28% are moderately impaired), and three standard deviations as *severely impaired* (0.13% are severely impaired).

Therefore, in terms of psychometric test reporting, based on that unique perspective and probable needs, a normal classification can include from 84 to 99% of the population. Of course, there should be considerable care in labeling the remainder as "impaired" to any degree, because of the social and personal consequences.

A human error perspective

In contrast, the human error specialist should understand that even those judged normal, on a behavioral attribute, could commit serious human errors.

Perhaps those that are impaired may commit a higher frequency of human error. The ruling assumption is that as each behavioral vector increases in symptomatic severity, there should be a commensurate countermeasure remedy.

This perspective is best illustrated by the following description of the underlying logic and dogma necessary for human error prevention.

Those involved in the design of consumer products often forget that half of the *general* population is, by definition, below average in intelligence. They are too often lacking in other desirable personality, behavioral, and physical attributes. This reveals much of the cause of human error, that too much may be expected of some subgroups. The ordinary consumer may become confused by even simple instructions, may become inattentive easily, may forget what he or she has done in the past, may misinterpret procedures, may become frustrated by minor complexity, and might ignore or even reverse the meaning of warnings. What may seem obvious to the designer of complex equipment may have become obvious because of some recent personal familiarity with the equipment or from past experiences of working with such equipment.

A *special user* population, if clearly defined and understood, may present few human error problems. Such limitations on use are difficult to maintain. Special user expectations should not be ignored, nor should their common practices. Even the highly intelligent user may not have the time or inclination to deal with unnecessary complexity, burdensome servicing, equipment installation, start-up, calibrations, preuse procedures, or difficult memory requirements. Some may avoid the performance of dirty tasks, routine maintenance, or servicing procedures that seem to them to be outside the confines of their assigned work or occupational specialty. What they expect and desire should be determined to avoid human error by sophisticated operators.

The general population, in current economics, includes almost anyone in the world. A special user population, when specified, may include a narrow population such as physicians and medical technicians using medical devices. It may be aircraft pilots using a display and control subsystem. It could include electrical engineers using test equipment for new software. It might be so narrow that only trained uniformed specialists may use monitoring equipment for radiological or nuclear systems.

A subgroup user population may be affected by a fairly common bias that affects how things are seen, perceived, or translated. The bias may originate in cultural norms, political affiliation, religious beliefs, or economic status. Each subgroup may have a tendency to respond differently to routine, repetitive, and monotonous tasks. They may respond differently to unfamiliar events, the need to interact or coordinate between individuals or other groups, or to take appropriate action as harm occurs.

Therefore, the perspective of the human error specialist must encompass a broad spectrum of knowledge relevant to determining cause and prevention. This includes foreseeable human behavior, including some understanding of the human functions identified in neuropsychological examinations (described below).

Neuropsychological test evaluation

The evolution of neuropsychological testing is impressive. Such tests have substantial credibility when strict protocols are used to administer each test, a carefully selected battery of tests are used to properly define multitest profiles, and licensed examiners evaluate the test results. The tests may identify specific brain functions and their associated neuroanatomical areas that could be impaired, function normally, or be above average. The tests correlate with clinically specific disease entities, neuroimaging findings, and neuroscience concepts. Neuropsychologists have earned the respect and general reliance often asserted by clinical neurologists in specifically identifying cognitive function impairment that might be manifested in closed-head brain trauma. Certainly, the test development process has a good foundation, the individual tests are vigorously and critically revised, and subgroup differences in age, sex, education, language, and culture are highlighted for assessment.

Such tests are far better than simple pen-and-pencil questionnaires dealing with a test author's concept of personality traits or other behavioral functions deemed to have some descriptive or predictive value. But 8 hours of testing on one individual is costly and time-consuming. It may constitute a violation of privacy, since personal, sensitive information may be subject to unauthorized release. Even with informed consent, there may be incidental findings of unexpected or previously undetected abnormalities.

Closely related are the ever more precise neuroimaging techniques that may supplement neuropsychological testing. Those brain scan techniques include functional magnetic resonance imaging (fMRI), positron emission tomography (PET), near-infrared spectroscopy, and other imaging devices. Various research centers have studied the relationship between the activation of certain brain areas and personality traits, risk aversion, deception, truthfulness, persistence, aggression, empathy, reasoning, negativism, anxiety, emotions, and other behavioral characteristics. There are general questions whether the brain imaging, neuropsychological testing, and neuroscience constructs yield a more scientific type of information. Do they suffer from the mere hope of explanatory depth, go beyond the limits of current understanding, or suggest too much of a phrenology approach and not enough about holistic neural circuit interactions throughout the brain? Is it just the beginning of neuropsychological application and usefulness?

It should be cautioned that the terminology and nosology are based on the clinical diagnosis and treatment of mental disorders, even though they are also used in educational and research endeavors. The widely used and internationally accepted standard is the *Diagnostic and Statistical Manual of Mental Disorders* (DSM-4), published by the American Psychiatric Association, in Arlington, VA, in 2000. A similar terminology and classification system is not yet available for human error purposes. The DSM-4-TR also presents a multiaxial assessment system where axis 1 includes clinical disorders, axis 2 personality disorders and mental retardation, axis 3 general medical conditions, axis 4 psychosocial and environmental problems, and axis 5 a global assessment of functioning. It is a biopsychosocial model of individual assessment. A similar model that has a focus on human error may result from factor analysis studies (Fruchter, 1954) and peer consensus, but is presently not available. Thus, the following clinical model is that which is available for predictive and research purposes.

Neuropsychological testing may be supplemented by an assessment of an individual's ethic and cultural identity and any possible psychosocial conflicts between the host culture and the culture of origin. If there are patterns of distress or troubles that are abnormal, the individual may be deemed culture-bound. The influence of local culture is important if it produces repetitive patterns of unusual behavior.

In summary, this type of testing may become an important tool in the search for a better understanding of human error, its causation, and its control. Because of its research potential, a brief description of some of the currently used neuropsychological tests is presented below.

Abstract reasoning

The *Wisconsin Card Sorting Test* (WCST) is commonly used to assess abstract thinking, perseveration, frontal lobe dysfunction, and executive functioning. It measures problem-solving strategy, including conceptual planning, organized searching, flexibility in regard to environmental feedback, the direction of behaviors oriented toward goal achievement under changing stimulus conditions, maintaining a cognitive set (maintenance of task set), and the modulation of impulsive responses. A computer version 4 (on CD-ROM) utilizes an on-screen presentation with audible messages in 10 languages, a keyboard or mouse entry, and software report generation. The raw scores convert to a normalized standard, percentile, and T-scores. There is a short version, the WCST-64, that can be administered in 15 minutes, as opposed to 30 minutes for the standard version of the test.

Memory

The *Rey–Osterreith Complex Figure* (ROCF) test measures the encoding of visual information, by figure copying, based on initial recall and the delayed recall (rate of forgetting). There is a 738-page handbook about the administration, scoring, and interpretation of this test.

The *Rey Complex Figure Test and Recognition Trial* (RCFT) measures visuospatial memory in terms of recall and recognition, as well as processing speed, response bias, and visuospatial construction ability.

The *Wechsler Memory Scale Revised* (WMS-3) provides information on working memory, recognition, immediate and delayed recall, retrieval deficits, and learning.

The *Wide Range Assessment of Memory and Learning* (WRAML-2) measures immediate and delayed memory, working memory, verbal memory, visual memory, and symbolic memory.

Cognitive processing

The *Stroop Color and Word Test* measures cognitive processing based on the fact that individuals can read words faster than naming colors. It measures coping ability by dealing with cognitive flexibility, creativity, and resistance to external inference.

Substance abuse

A substance abuse disorder may be identified by use of the 20-minute *Substance Abuse Subtle Screening Inventory* (SASS-3). A 15-minute Spanish SASSI version is also available.

In cocaine users, there may be cerebral vasoconstriction, pupil dilation, hypertension, perspiration, restlessness, hyperactivity, stereotyped behavior, euphoria, anxiety, grandiosity, and other symptoms. The neuropsychological tests may indicate an intact intellectual functioning, long-term memory, and verbal fluency. But they may also indicate an impairment in attention, concentration, mental flexibility, and visual learning. There may be cortical atrophy with prolonged cocaine use, with possible brain hemorrhages and seizures.

Alcohol, a central nervous system depressant, is in widespread use. Alcohol intoxication produces slurred speech, poor muscular coordination, eye nystagmus, increased aggression, mood changes, and short-term cognitive impairment. Chronic alcoholics may have deficits in problem solving, abstraction, perceptual motor ability, and nonverbal memory. Prolonged use may result in some cerebral atrophy and poor performance on neuropsychological tests sensitive to frontal lobe dysfunction.

Marijuana, also widely used, produces cognitive deficits in terms of problem solving, attention, language, memory recall, and executive functions.

Addiction involves an alteration in rational decision making (behavior choice), error correction (reward prediction), impulsivity (preference for immediate gratification), and, neurobiologically, a bias in the differential activation of dopamine neurons. From the human error viewpoint, these behaviors are undesirable, present learning and training problems, and supplement other associated clinical symptoms.

Intelligence

The *Wechsler Adult Intelligence Scale–Revised* (WAIS–R) provides an estimate of verbal IQ, performance IQ, and full-scale IQ based on a standardization by age, gender, race, geographic residency, urban–rural involvement, and educational level. Socioeconomic and demographic variables are also important. Since brain damage does not affect all subtests equally, the less vulnerable subtests are used to estimate premorbid intelligence. There is also an age deterioration index based on tests that tend to stay the same with age and those that decline more rapidly.

The *Reynolds Intellectual Assessment Scales* (RIAS) also measure general intelligence, including problem solving, verbal reasoning, and learning ability. The *Leiter International Performance Scale–Revised* (Leiter–R) is a nonverbal test of intelligence and cognitive abilities.

Aphasia

The *Boston Naming Test* (BNT) provides an assessment of object-naming ability useful in diagnosing aphasia (acquired language impairment). This test measures the ability to produce words given prompts, thus measuring language and communications ability. The *Boston Diagnostic Aphasia Examination* (BDAE), 3rd edition, measures the severity of impairments in communication skills, including word comprehension, word production, and reading disorders.

Malingering

The *Structural Inventory of Malingered Symptomatology* (SIMS) is used as a screening instrument to detect probable malingering, in conjunction with other tests. Malingering is considered to be a willful poor performance, a less than optional effort, the exaggeration of symptoms, or the intentional production of false information, usually in the pursuit of some external reward or to avoid responsibility. This negative response bias may be detected by such tests as the *Symptom Validity Test* (SVT) for blatant exaggerations, the *Hiscock and Hiscock Procedure* for memory bias, and the *Minnesota Multiphasic Personality Inventory* (MMPI-2) scores, which may raise a suspicion of malingering.

Cautions

1. There are many different tests, and each neuropsychologist may have different personal opinions, based on past experience, as to which combination is best suited for a particular individual.
2. There should be no dependence on any one test or subtest. The overall test results should be assembled, cross-checked, and integrated into comprehensive diagnoses and predictive treatment options.
3. Specific impairments should be traced to particular brain areas and interactions. For example, difficulty with words (anomia) may be traced back to angular gyrus lesions.
4. The implications derived for use by human error specialists, whether directly or by analogy, remain assumptions or hypotheses until proven otherwise.

Caveats

There should be an appreciation for the following nonconscious mental mechanisms as to their possible human error effect, possible neutralization or enhancement of countermeasures, and research possibilities regarding human error causation:

Every human error *choice* has emotional components derived from primitive analogies to gross situational responses.

Every human error *decision* is preceded by a preparatory and readiness mental state.

Every human error involves a *prediction* in the form of a likelihood assessment of the range of available alternative tasks and behaviors.

Every human error is based, to some degree, on current *beliefs* about past experiences that may be true, fragmentary, distorted, or unfounded.

Every human error *action* is based on mental perceptions of need, immediacy, relevancy, logic, and appropriateness under the circumstances.

Every human error is eventually *considered* and may be performed if it seems to have some desirable purpose at the moment.

chapter six

Countermeasures (remedies)

Introduction

A great deal of effort may be expended on human error detection, identification, delineation, and refinement, but no real correction of the error problem can be attained without some specific countermeasure to control, contain, reduce, or eliminate the risks. Throughout this book there are examples of human error problems and the various remedies that were or should have been utilized to minimize the risks. Those examples show that there are many countermeasures that could be selected and tailored to meet various specific needs in given sets of circumstances. This chapter will provide additional general principles and a virtual checklist of specific countermeasures.

General principles

The following 12 basic principles of error countermeasures serve to illustrate the intent, function, and general level of effectiveness of various remedies. They are arranged in descending order, from a high degree of effectiveness to virtually no constructive effect.

1. *Eliminate* the source of human error (remove the hazard). Make the error impossible by design.
2. *Control* the opportunity for error by physical means (engineering controls to prevent error, such as guards or barriers to prevent access to a source of error or hazard).
3. *Mitigate* the consequences of an error (risk severity reduction).
4. Ensure *detectability* of errors before damage occurs (foster immediate error correction).
5. Institute *procedural* pathways for guidance and to channel behavior (error avoidance by restrictions and narrowing of conduct).
6. Maintain *supervisory* control and monitoring for errors (error observation, oral directions, and manual shutdown). Particularly useful for new tasks, new employees, new jobs, and new equipment.

7. Provide *instructions* that are written, brief, specific, and immediately available. However, unsupervised employees do not always follow instructions.

8. Utilize *training* to provide general background information, job context, knowledge about company culture, and safety rules. However, training may be remote in time and not specific to a job task.

9. Have *technical manuals* available for reference when questions arise or for general self-learning. They are useful to help avoid trouble-shooting errors.

10. *Warnings* provide an informed opportunity to avoid harm. They are used for residual risks after the use of other remedies. Effective if well designed, read or heard, understood, and not disregarded.

11. Specify that *personal protective equipment* or other safety equipment be available when and where needed (for harm or injury reduction).

12. Assume intentional *risk acceptance* by no error prevention action and a toleration of the results. Provide appropriate insurance coverage or company reserves to compensate for the foreseeable harm. Maintain recall and public relations plans.

There may be an inherent difficulty in implementing countermeasures if there is a strong belief that the worker, user, or consumer is fundamentally responsible for his or her own error-free performance and that he or she should fully accept the consequences of his or her own actions. This conviction about an overriding personal responsibility may be accompanied by ideas of an implied voluntary personal assumption of any risks associated with work tasks or other activities. In addition, there may be a notion that every person facing a task is basically the same as a reasonable, average, prudent person. There may be another notion that that person will face a task with a tabula rasa mind-set devoid of the influence of past habits, customs, beliefs, and experiences. In addition, there may be pervasive opinions as to probable exemptions from regulatory reporting and enforcement actions; exclusions by virtue of developmental, experimental, limited use, or state-of-the-art status; or a lack of proof as to hazard causation and ambiguity as to the ultimate costs of any harm created. In other words, countermeasures intended to prevent error or fault should be selected and devised by persons who do not have some philosophical or expert bias that could distort what is actually needed, available, and could prove effective.

Specific countermeasures

Single-error tolerance

It is a cardinal principle that a single human error should not result in appreciable harm to persons or property. The product, machine, or equipment should tolerate a simple mistake, an inadvertent goof, or an honest error without giving rise to a dangerous condition. A single human error includes repeated mistakes of the same type being used in the attempt to achieve one objective. A distinctive feature of human intelligence is its flexibility in adapting to novel, changing, or complex social and occupational situations. Exploratory behavior, when faced with uncertainty, is a trial-and-error learning process. False beliefs (mistaken perceptions) result from that mental flexibility and a constructive search for correct answers, relevant information, plus subjective interpretations by the brain based on the activation of encoded frames of reference (in both the active prefrontal cortex and the altered thresholds in nonfrontal regions of the brain). A false belief can result in erroneous human decisions (errors); then a mental correction process can occur both to the frames of reference and to changes of beliefs (a learning process), and that may result in altered expectations and more appropriate error-free responses. Thus, tolerance of errors is a necessity to permit human learning and to accommodate foreseeable error in that process. *Caution*: A single human error may be a precursor or antecedent to other related errors that could occur simultaneously or successively. Thus, apparently single human errors, causing no immediate harm, should be carefully analyzed for related mechanisms of harm.

Human fault tolerance means that a system will continue to operate despite human error, loss, or failure. In essence, the system tolerates human error without a loss of normal function for some designated period. The error might be considered a transient glitch to be overridden or smoothed out by the machine, an error of omission or commission to be self-corrected by a computer, or an error in a smart system that can independently exercise its own control where mistakes occur in time-triggered sequences of synchronized events. Built-in error override, self-correction, and retained control are important for safety-critical functions.

The rule of two

Design engineers should consider occupational safety and health when designing equipment. For example, an instructional book for mechanical engineering students with little exposure to "the fundamental concepts of occupational safety and health" indicated that injuries could result from human error (such as worker misjudgment or omission) or design error (such as brake failure), and the task of the designer is to prevent adverse results due to such errors (ASME, 1984). This textbook supplement described the "design rule" as requiring at least two independent human errors or one error and one independent equipment malfunction to avoid "dangerous operating conditions." If the results could be catastrophic, the two error rule should be increased to three or more independently occurring errors before a danger ensues.

This instructional aid cautioned against trade-offs between safety and other design parameters and urged that design reviews for safety be undertaken independently of all other design considerations. It suggested that job-related stresses could prompt workers to pay less attention, tolerate risks, and take shortcuts that could increase their chances of becoming injured on the job. That is, the designer should consider all the risk factors that could increase the propensity or likelihood of human error.

Interposition

If there is a recalcitrant human error, it may be necessary to interpose a barrier, shield, shroud, or to otherwise deny access to a hazard or potential source of harm. Interposition functions as a pathway interruption or breach of causation. Examples include point-of-operation guards, barricades, electrical insulation, covers, screens, fencing, shielded fuel tanks, truck cab guards, presence-sensing devices for deactivators, and computer firewalls. It is access denied, to both hidden and obvious hazards, from undesired human behavior. But guards are often removed, forgotten, or replaced incorrectly. They may need permanent attachment, interlocking, or to be supplemented with a jog (limited movement) function to enable troubleshooting, maintenance, servicing, or repair operations. The guard may protect people from injuries or machines from damage. Interposition simply separates the error from the source of harm, by interposing a barrier between them.

Sequestration

It may be desirable to completely isolate the error or the source of harm. It may be by providing a sufficient distance between them — something beyond the human reach distance or a location for the source of harm that is separated, secluded, detached, segregated, or set apart. Examples include a remote flammable gas area, an off-limits hazardous waste dump, a posted no-access contaminated or polluted site, the marked limits of a munitions storage area, an enclosed or normally remote railroad third-rail location, a cordoned area used for abrasive sandblasting or painting, a restricted use storage area, or a demarked or colored electrical leakage zone or boundary. Physical sequestration removes the uncertainty of whether errors can be eliminated by due care, proper supervision, clarity of procedures, and enforcement of work rules.

Interlocks and lockouts

A worker may need to perform cleaning, servicing, maintenance, or repair tasks on a machine. In the process, he may move into a visual blind zone located behind, under, or within the machine. The machine operator may not see him, mistakenly turn on the machine, and crush or otherwise injure the maintenance worker. To prevent this fairly common scenario, there is a need for *lockout* procedures and devices. The power switch, on the wall or on the machine, should be locked in the off position before work starts. The common practice is to use a padlock for each worker, who retains the one key that fits his lock, and then tagging the power switch lockout with his name, department, and description of the work he is performing.

This may be supplemented by the use of *locking pins* to prevent shaft rotation and *lock bars* to prevent closing dies and platens or movements of structural elements of the machine.

Access doors for maintenance operations, which cover or hide hazards, should remain locked until the machine stops running and there are no exposed moving parts. This can be accomplished by using door *interlocks* where the door remains locked until the hazard ceases to exist.

Interlocks can also provide event sequencing that will control potentially unsafe behavior. They function as a deactivation device to prevent harm. Examples include an access door switch that, when the door is opened, cuts off the electrical power so there are no exposed powered electrical contacts. A prescribed sequence of events may be required before the final lockout (interlock) event takes place. The lockout prevents improper, inadvertent, unwanted, or untimely behavior to actuate equipment. An interlock may prevent an automobile navigator, information display, or other distractors from functioning unless the vehicle is stopped or the display is pointed away from the vehicle driver.

When stopping occurs, it should be a power removal to all actuators. It can be a controlled stop or a fast-braked stop. The machine should not restart while a potentially dangerous condition exists, such as an open access door. The possibility of a failed relay, stuck in the closed position (not a fail-safe condition), may require circuit redundancy, monitoring, reset inhibition, or other design safeguards.

Channelization

Errors may be reduced by providing guidance information to facilitate appropriate choice behavior. The automobile driving task is guided or channelized by roadway lane markings, speed zones, directional signs, traffic lights, road edge marking, curve delineators, raised pavement markers (Botts Dots and RPMs), highway lighting, deceleration lanes, emergency parking, crash cushion devices, and guide signs. Their location may be determined by the 85th percentile vehicle speed and the PIEV time (perception, identification, understanding, emotion, decision making, and volitional execution time, which ranges from 3 to 10 seconds). The result is to channelize the mixed traffic flow in a manner that reduces accidents for all roadway users, pedestrians, and bystanders. The information that is presented should be clear and positive, uniform, command attention, and be available when needed. This merely illustrates that human behavior can be channelized in complex circumstances to avoid confusion, disarray, disorder, mistakes, and errors of the decision-making process.

Guides and stops

A piece of material passes through a woodworking machine and comes out of alignment; the operator inappropriately, but with good intentions, reaches in to reposition the piece, but his hand inadvertently contacts the rotating cutting head of the machine. A component part of a product gets stuck on a conveyor, the operator attempts to loosen it, and his hand jams in the machine. A baggage handling system at an airport is abandoned because of damaged, misdirected, and lost luggage. These problems illustrate the need for physical guides that serve to counter the foreseeable misalignments, jams, and damage that can induce attempted human rescue errors.

A machine operator feeds a large piece of sheet metal part into a press brake to bend it; however, he pushes it too far. There is no stop device on the rear of the machine to help accurately position or locate the metal sheet. His error is easily corrected by installing a device to stop the metal sheet in the correct position. A similar problem exists on some powered saws, where there is no right-side alignment guide (adjustable bar or rail) and no rear stop device (adjustable metal fingers).

Sometimes there are very simple remedies that can help a machine operator do his job. There should be no perceived challenge to the skill demanded for machine operation, since the challenge is error reduction.

Automation

One of the most common responses to the discovery of a significant human error problem is a quick retort to take the human "out of the loop" or to try to completely remove the human error source from the system. This is not easily accomplished and may be cost-prohibitive. Humans tend to be low cost when adaptation to frequent change is necessary. Most important is that automation shifts the errors from manual machine operation to automated machine servicing, maintenance, adjustment, troubleshooting, and repair. There are robot material transfer machines. There are computer-controlled machine tools that perform a number of metalworking functions, have built-in inspection capability, and can automatically replace tool heads that chatter from wear. There have been completely automated, virtually worker-free factories. There is certainly a trend toward automation, but it may not be an immediately appropriate remedy for specific human errors.

Instructions

When there are no specific, detailed, easy-to-follow directions, human error abounds. Instructions provide knowledge about the operation of equipment, explain correct procedures, and assist persons in achieving desired human performance. They may be needed to help prevent unnecessary and preventable harm. They should be in a form that is simple and straightforward, easily accessed, and available where and when needed, and they should provide helpful information. Examples include service manuals for technicians, owner's manuals for consumers, and troubleshooting guides for service engineers. Instructions should help identify when task demands could exceed intrinsic human capability to facilitate avoidance behavior without error or harm. There is a trend toward providing up-to-date instructions by computer disks or by the Internet for personal computer or television display. To avoid complex words and obscure symbols, instructions should be tested for a specific grade level, literacy, and ease of readability by tests such as Flesch-Kincaid, Coleman-Liau, or Bormuth. Instructions may be considered a continuing education form of training.

Training

A so-called commonsense reaction to disturbing allegations of human error is to write on an incident investigation form that more training is necessary. It is an unfortunate reality that most unskilled workers, factory technicians, and engineers receive virtually no training upon employment, subsequent transfer to a new job, or at the time of an assignment to a new project. Training costs money, time, and diversion of talent. The most common answer to questions about training is that on-the-job training was received, which generally means undirected self-help during job performance.

Training means to instruct or guide in a designated or formal manner. It functions to provide certain basic skills or essential knowledge by means of demonstration, participation, lectures, printed manuals, specified curricula, and appropriate learning processes. Training may be useful for a general familiarization with a complex system or for the imposition of prescribed standard work practices. Training may be costly since its effects may be incomplete and transitory. The primary problem is that training may be considered as a wastebasket remedy when other countermeasures cannot be determined or implemented. When training is designated as a countermeasure, it should be understood that there are many different types of training, and each may have a different result. Training may last from 1 hour to weeks, it may be unsophisticated or highly tailored to precise objectives, it may be relegated to an improvised self-learning exercise, or it could be conducted on specialized simulators in special training facilities.

Training or retraining is a common prescription for skill reinforcement or where habitual patterns of behavior need to be replaced with something new. A forced intentional learning situation is far more effective than an optimistic hope of self-improvement where the need is for more immediate error-free future performance. The use of problem-solving conferences could lead to error correction information not otherwise available. Retraining specialists may assume that reduced error rates could increase productivity and that the resulting economic gains are the best way to win management support for more widespread training efforts.

There may be mandatory government-required training to ensure some familiarity with material safety data sheets (chemical hazards), closed-space entry, and other workplace hazards. This may be in the form of a brief reminder at a tailgate session or a more formal periodic refresher training. This is generally restricted to certain workplace employees and limited by the competency of the assigned instructor.

The example of astronaut training illustrates one kind of careful selection of trainees, their intensive training, and their preparation for all probable misfortunes. This form of training is costly and time-consuming, but effective under the circumstances. It is in stark contrast with the supplier of domestic kitchen appliances, where formal training is all but impossible. The household appliance must be intuitive in operation, that is, an operation capable of quick apprehension or knowing by the user without recourse to written instructions, mental reasoning, or the assumption of logical inferences. It should also be capable of preventing error that could have adverse consequences. This is the general consumer expectation: operate the device or machine without the need for training.

Behavior modification

There are advocates for what is called behavior modification and attitude adjustment. This approach may entail interventions to improve job satisfaction, interpersonal relationships, and company cultures. Generally, it involves observing what people do and getting them to think and act differently about at-risk behaviors. Desirable behavior is achieved by positive motivations rather than punishment. Some of these advocates decry prediction of behavior on a lifestyle or personality basis. They attempt to motivate safe work habits. Some of the advocates stress management conviction and measuring various indicators of safety activities within a company. If classified as a motivational program, behavior modification bears a resemblance to many quality assurance programs to reduce manufacturing discrepancies and defects. It is assumed that human error in the workplace can be significantly reduced by such activities. The results of such programs vary, and they may or may not have lasting effects.

Safety factors

Safety factors are used, in design, to accommodate the unknowns and uncertainties by some compensation or seeming overdesign. A conventional illustration is mean *strength* divided by actual *load*, yielding a safety factor ratio that can be increased where human life is concerned. Adequate safety factors are a precautionary measure. A material failure during test may require increased design strength, as a countermeasure to the discovered weakness or propensity to fail under load.

An illustration of the use of safety factors, in the strength of materials, is the recommendation to use a safety factor of 1.3 to 1.5 for highly reliable materials, 2.0 to 2.5 for ordinary materials, and 3.0 to 4.0 for materials that are not reliable or used under difficult situations (Oberg and Jones, 1973). Another source recommends a safety factor of 3 for known loads where there is no metal corrosion, 5 or 6 for alternating loads, 10 where there are repeated shocks, up to 40 where complex strains may occur, and even higher when failure could result in loss of life (Carmichael, 1950). *Caution*: When used to determine allowable loads, a safety factor of 3 does not mean that the load that can be carried is up to three times the load for which it is designed. The safety factor is merely an allowance for the uncertainties of actual use. Failure can occur before the load suggested by the safety factor is reached.

In human factors engineering, a safety factor provides specified allowances to compensate for the uncertain range of individual human differences, given the circumstances, equipment, and working conditions under consideration. It could be expressed as the ratio of ultimate human capability, propensity, or capacity to the actual on-site work demands or performance. It could be the strength divided by the stressors. The need may be for a higher or lower numerical safety factor given the expected or predicted variance in the desired human performance. It is a shorthand means of describing the accommodation of all reasonable variations in human behavior beyond that normally expected.

In terms of human error, the applicable safety factor may be the ratio between the level of errors that can be *accommodated* by the design or situation, as compared to (divided by) that which is reasonably *expected*, anticipated or foreseeable. It could be a ratio of variances. It could use the expected residual error (from risk assessments). It could use the lifetime error expectancies (total exposure).

One important question is whether the *actual* error load (reasonable expectations or design intent) includes an extreme error load that is of low probability yet a foreseeable event. If the consequences of a human error include catastrophic harm, a relatively high safety factor for error is desirable as a preventive measure to reflect the uncertaintes of the real world.

This is not a recognition of the old maxim that if anything wrong can be done, it will be done. It has to do with accounting for the variations normally expected in human performance and also the uncertainties surrounding human error. It is far better to have an adequate safety factor or safety margin to rule out surprises from human error.

Caution: Although the basic concept of safety factors is widely used, there are many debatable interpretations, including what factors or conditions are considered uncertain and what are foreseeable. Estimates of human capability normally include statistical variance, but uncertainty is more than simply accounting for the expected. Human error safety factors are independent of safety factors used by other disciplines.

Warnings

Some degree of error avoidance may be achieved by the use of warnings for residual risks. They can alert a user to hazards that could injure the person or damage the product, process, or equipment. The purpose is to enable avoidance behavior. It is a form of informed consent where, once informed of the hazard and risk, the individual has a choice either to avoid or to voluntarily assume the risk. Warnings are used for residual risks that cannot be reasonably minimized because of product development schedule timing, technical hardware complications, excessive economic burdens, or other practical considerations (Peters and Peters, 1999). Warnings may be in printed form, such as in chemical hazard labels, recall notices, or safety bulletins. They may be in visual form, such as a blinking red air bag warning light or some reflective pavement markers. They may be in tactile form, such as roadway speed bumps, rumble strips, or aircraft stick shakers. They may be auditory devices, such as grade crossing alarms, backup beepers, horns, sirens, whistles, or automated voice commands. They are found everywhere; for example, crane work platforms may have slope or tilt alarms, there are aircraft voice warnings, and automobile low-fuel indicators. Each has an injury history (need), evolving technical requirements (specific solutions), and some degree of sophistication (a level of effectiveness).

Implanted defibrillator devices, for treating patients with congestive heart failure, have suffered electronic malfunction and have required a voluntary recall or replacement. The failure may be an electronic switch stuck in the open position so the needed jolt of electricity to the heart is not delivered. But patients have been warned by an audible beeping sound so that there will be no mistaken reliance (error) on the device. The benefit of these devices is that many of the 400,000 Americans that die of sudden cardiac arrest could be saved by implanted defibrillators (Burton and Pulls, 2005).

Protective equipment

There may be airborne contaminants near a work station in a factory. They may be from welding gases, lift trucks with internal combustion engines, mixers of batches of toxic chemicals, leaks and spills of hydraulic and lubricating oils, and metal dusts created from grinding operations. While there may be an obvious health hazard to the trained professional, there may be some more immediate problems, such as lack of coordination, nausea, vertigo, and other central nervous system effects. There may be personality changes, errors of judgment, and errors in how work is performed. The countermeasure to questionable amounts of toxic materials is the use of respirators, sleeves and gloves, and protective clothing. Equally important protective equipment includes detection instruments with alarms, fire protection devices, negative-pressure glove boxes, ear protection, vertical-flow hoods, air locks, pass-through ovens, static control devices, and local exhaust systems. In other words, the hazards that can induce human errors may be controlled by personal protective measures. However, such risk reduction is better accomplished by appropriate initial design remedies.

Redundancy

A system failure may be caused by human error or the failure might provoke subsequent human error. Redundancy simply means that the design should provide more than one means to accomplish a task, a given function, or an ultimate objective. It may be an alternative, extra, standby, or duplicative element where system failure could have catastrophic results. It may be used where there is a suspect weakness in terms of possible human error, a probable lack of immediate supervision, or a strong need for a robust design. Examples include twin taillights for automotive vehicles, multiple truck reflectors, more than one engine on aircraft for passenger safety, and the use of several operators for overlapping or pass-off display and control functions. Human redundancy could be simply having a two-man team, an operator and a checker, to doubly ensure a critical task is accomplished correctly.

Derating

Errors may be reduced simply by reducing the workload on the human operator. It serves to reduce the possibility of a human overload that can result in confusion, mistakes, or sudden drops in the desired human performance. Derating may be advisable where several additive loading factors could be present, some emergency or panic situations may occur, or the safety factors are not adequate (see Safety Factors section above). Humans are essentially single-channel information processors; therefore, they are susceptible to information overload in terms of time to respond, correct choices, and manual output. Derating simply means requiring human behavior below that which the operator is capable of performing.

Fail-safe

Equipment should fail in a manner that creates no significant hazard. This means that a human error causing equipment failure should not create an unacceptable risk. This may be accomplished by ensuring that an error is immediately correctable, harmless, or one that will be ignored by the system as an expected perturbation. An example is the railroad engineer dead-man switch that will apply the brakes if the operator fails to maintain a required force on the safety bar. He can correct the brake application if it is an error. If he has a heart attack or drops dead, the train will come to a safe stop. Another example is the run-flat tire. When a slow leak, puncture, or blowout causes tire deflation, the vehicle will not go out of control, can be driven to the side of the road or for some distance for assistance (repair or replacement), and the occupants avoid the dangers of roadside distress. Efforts should be made to ensure that errors causing equipment failure result in a fail-safe situation.

Stress reduction

A high level of prolonged stress will produce physiological changes in the human body that will increase the likelihood of errors of judgment or operational mistakes. A good example of a job that creates a high level of mental stress is the airport air traffic controller. For hours at a time, the controller must interpret radar indicators of moving aircraft that change speed, direction, and altitude in relation to other traffic patterns, traffic densities, pilot intentions, available airspace, and airports with incoming and outgoing flights. The controller knows that an aircraft collision involves the lives of many passengers, the flight crews, and those on the ground at a crash site. To prevent such catastrophic human error, the controllers are carefully selected, trained, and certified. Near collisions are carefully studied (Meckler, 2005) by facility managers, inspector general personnel, and research scientists. But there are continuing incidents of airplanes flying too closely, aircraft takeoffs with other aircraft on the same runway or crossing it, and other near-miss events. Some controllers have been punished by probation or decertification (requiring retraining), and there may be ground changes such as signs and warnings. There is a need for controllers who can manage stress and feel comfortable with high stress levels, remain calm but alert, repress emotions in an emergency, and have a willingness to disclose what others believe are errors.

For many tasks, it is generally believed that job stress results in tension, anxiety, fatigue, high blood pressure, and disruption of the regulatory functions of the human body. These effects or symptoms have been linked to judgment error problems and poor overall human performance. In other words, ordinary stress can result in increased human errors. Heat stress can kill. Industrial hygienists have been concerned with chemical, ergonomic, emotional, environmental, and physical stresses. The assumption is that if stress is a cause of human error, then the appropriate countermeasures to such errors are those ways in which stress can be limited, controlled, and reduced to optimum levels. For example, in the working environment, there should be some control over extremes of temperature, pressure, vibration, radiation, noise, repeated motions, biological agents, infectious diseases, excessive workloads, demands for unusual accuracy, and other factors that could contribute to stress reactions.

Tools

During error causation interviews with individuals or small groups of workers in a manufacturing facility, it is often surprising what is said about the tools needed to accomplish their assigned tasks. They are the persons attempting to do a good job, in an efficient manner, and without errors, so they have direct nontheoretical knowledge as what is actually needed. They may be frustrated with the tools they have been given, concerned about the delays in getting something better, or puzzled by their insight as contrasted with the indifference of their supervision. To avoid error and discrepant parts, there should be appropriate measurement and inspection tools. A bad indicator is the presence of mushroom-headed chisels, cracked wrenches, dull cutters, hammers with loose heads, knives without antislip handles, and electrical power tools without double insulation. Bad tools can induce human errors that could affect the final quality of a product. During the initial design of a product, the needed tools should be considered in relation to what would be available, how the tools would be used, what could be done to avoid errors, and what would be required for special tamper-proof and security features, accessory devices, or repair operations.

Replacement

There are situations in which some humans persist in committing errors regardless of the consequences and despite the efforts by others to convince them that they should take some personal responsibility. They may say they forgot, were busy doing something else, or give other excuses in a relatively unapologetic manner. Others seem to enjoy the attention they receive for acts that run counter to the social norm. There may be behavioral disorders that could result in a repetition of errors (see Conduct Clusters section in Chapter 9). There may be a use of illegal drugs or other off-the-job causes of recurring unacceptable errors. If all attempted countermeasures prove unequal to the task, at some point in time the person must be replaced by job reassignment or termination.

Enhancement

The human operator may need enhancement, leverage, or other modification of his basic capabilities. For example, the force applied by the human foot on the automobile brake pedal is amplified and distributed to the brake pads on all four wheels, with a tactile (feel-type) feedback to the driver's foot. Similarly, there may be a variable force and distance actuation in the power steering of an automobile. Night vision is improved by headlights or, in military operations, by light enhancement devices. Hearing can be assisted by voice amplification and distant communications by radio or television sound adjustments. The force applied can be cushioned (for force impact protection), made proportional to movement (with force feedback signals), or countered to be made resistant to inadvertent actuation (error prevention). Differences in force feel may signal error or nonerror. Human intelligence can discern and interpret a wide variety of augmented (encoded) signals.

Inactivity

When a machine or product is turned off and the operator or consumer does not intend to continue its use, there is a reasonable belief that an inert or safe condition ensues. This means no electrical current or voltage flows through the machine or product during disuse, after being turned off, other than harmless trickle currents in isolated safety (self-check) circuits. There should be no reasonable probability of harm from electrical shock or ignition of combustible materials by a machine believed to be inert, powerless, or inactive. It is a design error, not a user error, if harm should result.

Independent confirmation

In the U.S., in the second case of mad cow disease (brain-wasting disease or bovine spongiform encephalopathy), there was a 7-month delay in alerting consumers. This "human error" was caused by "missing paperwork" and "sloppy handling" of the cow's brain sample (Kilman, 2005). An early brain test had cleared the animal, although an experimental test had detected abnormalities. Much later, the USDA inspector general's office decided to use a different group of scientists to test the brain samples. It was found positive and then sent to England for test confirmation that there was an early misdiagnosis. It is important because some consumers of infected meat can manifest the always fatal prion-caused brain disorder. Given the seriousness of such a health threat, the countermeasures to conflicting opinions and palpable excuses would seem to be a second opinion by independents and also a third opinion confirmation by still another independent test.

In essence, reversing a decision generally requires more than one opposite opinion to tip the scales of credibility when bias might exist.

Therapy

Errors may be traced to high-value employees who are virtually irreplaceable and give the impression that they can correct any undesired behavior themselves. This may be possible with some mentoring. It may require role-playing to help provide a better understanding of the situation and the consequences of various actions. Professional counseling may be helpful. However, intensive therapy may be excessively costly and time-consuming, with uncertain results and possible side effects. Individual and group therapy are options that should be rarely exercised within the company enterprise. Therapy may be considered a long-term exorcism that is a private matter for the individual as part of his own self-development on his own time. Therapy remains an option that has been used in the form of supervisory counseling for a general propensity to commit managerial errors.

Improvisation

There are situations in which extemporaneous remedies are fashioned, at the scene, to correct ongoing human errors. Examples include factory floor improvisation at the time the problems are encountered. The remedies may be comparatively simple, such as shadow boards that have outlines or shadows of various tools, to indicate what tool belongs in a particular space and whether it is present or missing. This helps to prevent tool crib errors. Improvisation is implicit in continuous, nonending process improvement efforts (such as Kaizen programs). Kits (a box of parts and tools) may be furnished to assemblers at the point of use (thus, no visits to the tool crib). But the kits will change with product and process improvements, including error correction changes. Error-provocative situations may become a problem only after a change in workers or work orders on an established assembly line.

The remedies may be more complicated, yet improvised to correct human errors. For example, in the manufacturing process, there may be some mistake-proofing devices (from poka yoke efforts) to prevent incorrect parts from being assembled (baka yoke), to identify flaws (errors), and to give warning of the flaws. When a flaw or defect is discovered, its production may be stopped until the flaw is corrected (jidohka). Similarly, a machine tool may automatically inspect each item, and if a defect is found, the machine will stop and call for human intervention (autonomation). The names given the error process help to focus attention on error detection and correction.

Improvisations in manufacturing operations may result from the efforts of industrial engineers or from labor union representatives having some responsibility for the health, safety, and productivity of workers. For quality assurance specialists, the first-pass yield, which is incoming parts less defectives, provides a percentage rating for scraps, reruns, and retested, repaired, returned, or otherwise diverted defective units. This could be a form of human error rating. Other indications of possible human error are accident incident, return-to-work, and light-duty rates. In retail distribution systems, there may be a chargeback system for the retail outlets. This is a financial set-aside for each lost time incident, which comes out of the bottom line of the store or distributor. All such indicators or set-asides serve to promote improvised corrective action. In other words, specific countermeasures may be the result of improvisation by those most directly involved. The improvisation indicates a problem and a solution that may be effective to some degree.

Note: There are many other safeguards, remedies, or countermeasures that could be considered where harmful error, mistakes, or goofs could occur or be induced. For example, there are emergency stop cables for conveyors to help prevent error entanglement injuries, there are pull cords to stop assembly lines to help prevent error omissions and discrepancies, and there are emergency stop buttons on some machines to help prevent unsafe acts or further injury on operating machinery. There are dual-input devices such as circuits on safety gates that require both "closed" and "not open" position signals prior to start or restart of a machine. There are limited inch or pulse timers, with buttons (switches) that must be released and reactivated for each movement of a machine that might have rollers that need daily cleaning. There are speed-limiting devices to prevent overspeed. There are light curtain devices to help prevent perimeter incursion, for stop–start operations, and to ensure proper material insertion. There are remote control devices for safe-distance operations. All of this indicates that the creative selection, design, and application of countermeasures is a professional-level activity of great social responsibility.

Selected countermeasures should be carefully discussed with system safety, reliability engineering, quality assurance, occupational safety, human factors, and industrial hygiene specialists where appropriate. They may have personal knowledge about such remedies and know how they could be quickly implemented, tweaked in some fashion, made more compatible, or somehow adapted to the circumstances. If no remedy has been found, search relevant trade standards, engage in Internet listserv discussion groups, or review textbooks and journal articles in reliability or other subjects that seem appropriate. Background information facilitates creative solutions to human error.

Pragmatism

There have been many attempts to categorize and name various types of human error with limited success. Some were a classification for descriptive scoring only, that is, a good or bad tally with a face value objective. Some were compilations of violations of work rules, regulations, or desired human performances, that is, a good or bad tally with a superficial improvement objective. Some were measurements of the dependent variable in experimental research, that is, the error response to a defined stimulus and having a restricted objective. These limitations can be overcome by adopting a more useful pragmatic preventive approach to human error.

The classification of human errors should be behavior based. Each category name should be descriptive of what has occurred in terms of human function. The categories should discriminate among the various types of human error. They should provide for root cause analysis and delineation of subtypes. Each category should imply or lead to preventive action. The criterion of this pragmatism is the usefulness of its consequences, improvements that reduce human error, a utilitarian goal.

Caveats

Targeting

Countermeasures may be targeted at or focused on the *human error* itself, such as error causation, its manifestation, muting, control, or elimination. It could target *causation*, the direct connection between the error and the harm, such as error detection and the isolation or separation of the error from whatever may be adversely affected by such an error. In addition, there might be a counterattack on the *consequences* or the harm itself, such as limiting possible damages and mitigating injuries. In essence, there may be more than one way to counter the harm from human errors.

Acceptability

A selected prime remedy may not be acceptable to others because of its cost, complexity, effectiveness, or a presumed interference with other design or marketing features. Be prepared to have second and third choices.

Multiple remedies

There may be a need for multiple remedies for one type of error. For example, consumer household table saws caused many finger amputations and other injuries because of hand-feeding errors or inadvertent hand movements near the unguarded saw blade. Then a cover or hood was used to guard the rotating circular saw blade on all sides. It significantly reduced injuries, but a second device, such as a fence guide requiring that both hands be well away from the cutting zone, was deemed necessary to further reduce the injuries. That is, two guards were necessary for adequate risk reduction because of the type of human errors involved.

Success metric

The success of an error specialist can be measured by how many improvements have been made. This means appropriate error identification, application of countermeasures that are effective, and the risk reduction actually achieved and measured.

Fade-out

As time passes, familiarity breeds complacency. With a more relaxed attitude, error-producing events just seem to fade away and become less important in job consciousness. Fade-out is not age dependent or error dependent. This form of forgetful indifference or fade-out of awareness can become contagious in the workplace by example or emulation to others. Forgetfulness could be countered by reminders that are novel, periodic, authoritative, and to the point, or by other, more lasting countermeasures.

Perversity

Adaptive behavior responding to change is a learning process where the physical substrate in the brain involves synaptic plasticity that is sensitive to small perturbations. Impaired learning and memory defects may result in normal workers from a lack of robustness, direction, and relevancy of the training process. This suggests that training given by amateurs may have some desired beneficial effects, some unwanted perverse effects, or both. If the error prevention potential of training is important enough, it should be accomplished in a professional manner.

Customization

There are numerous countermeasures available, as indicated in this chapter and throughout the book. They range from temporary patches to permanent fixes that will endure throughout the service life and afterlife of a product, system, or workplace. A thoughtfully tailored remedy appropriate to the life cycle is desired rather than some short-duration countermeasures.

Classification of human error

The names and categories chosen to describe human error should have the implied function of helping to prevent human error and the pragmatic goal of the improvement of work tasks, products, systems, and processes. The classification of human errors should be behavior based. Each category name should be descriptive of what has occurred in terms of a human function. The categories should logically discriminate among the various types of a human error. They should provide for root cause analysis and delineation of subtypes. Each category should infer or lead to preventive action. The criterion of this pragmatism is the usefulness of its consequences and the utilitarian improvements that help reduce human error.

chapter seven

Human factors design guidelines

Introduction

The study of human factors emerged from man–machine problems encountered during the Second World War. For example, aircraft pilots were misreading visual displays such as altimeters, selecting the wrong control knob, and interpreting navigation charts incorrectly. The problems were centered on the interface between man and the systems he was operating. This field of activity was called engineering psychology when there was a psychology emphasis or human engineering when there was a design engineering emphasis. As these activities produced substantial results in terms of product improvement, the discipline grew and prospered.

There are now 5000 human factors specialists, and there is academic coursework leading to doctoral degrees, many textbooks (Chapanis et al., 1949; McCormick, 1957; Woodson and Conover, 1964; Van Cott and Kinkade, 1972; Sanders and McCormick, 1997), and an impressive accumulation of peer-reviewed articles during the past 45 years printed in *Human Factors, the Journal of the Human Factors and Ergonomics Society.* There are many guidance documents, such as the standard practice (ASTM 1166-88), which established general human engineering design criteria for marine vessels (shipbuilding). A second standard, the marine system practice (ASTM 1337-91), established the requirements for applying human engineering in an integrated system engineering, development, and test program. A model specification (SNAME 4-22) for human engineering was published for used by naval architects. Anthropometric data (human dimensions) appeared in architectural design handbooks (Watson, 1997) and automotive vehicle design publications (Snyder, 1977). A human reliability handbook was published by the U.S Nuclear Regulatory Commission (Swain and Guttmann, 1980). This is illustrative of the widespread applications of human factors information.

Human engineering has been defined as "a specialized engineering discipline within the area of human factors that applies scientific knowledge of human physiological and psychological capabilities and limitations to the design of hardware to achieve an effective man–machine integration" (ASTM 1166-88). The term *human factors* has a broader meaning in terms of its objectives, such as enhanced human performance, human comfort, safety, health, productivity, psychosocial factors, and even quality of life.

Human factors involves a neverending search for a better understanding of just how humans interact with ever more intricate technology, in complex social situations involving diverse cultures and value systems. It would seem rather obvious that good human factors applications could, directly or indirectly, reduce human error and mitigate its consequences.

A significant expansion of human factors occurred with the formulation of macroergonomics, a sociotechnical system approach to the design of work systems and the harmonization of jobs, machines, and software interfaces (Hendrick and Kleiner, 2001). Any meaningful expansion of a professional discipline is a difficult matter, but macroergonomics was initiated, conceptualized, and fostered by Dr. Hal Hendrick, the former dean of the College of Systems Science of the University of Denver. This suggests that strongly motivated individuals can have a major effect on an emerging beneficial professional discipline, particularly when enabled by computer-processed bulk information gathered by digital sensor systems, videos, and other imaging processes.

In contrast to the application, practitioner, or design orientation of most human factors efforts, there have been attempts to better conceptualize, theorize, and develop models of human behavior in a systems context. There are attempts to deconstruct and reconstruct the facts and opinions relating to major disasters to form new theories of accident prevention. There are attempts to be less mechanistic or pragmatic and have much broader perspectives or a qualitative bias. As in the past, the future challenges presented may require new directions and approaches to human factors problems.

Methodology

The methods and techniques utilized by human factors specialists are quite varied and usually tailored to the need, the funds available, and any time constraints.

Scientific research

There has been an emphasis on experimental research in order to prove the objectivity and scientific basis of human factors. This is laboratory research in which all variables are controlled (held constant) except for the experimental variable (stimulus), which is manipulated to determine what happens to the dependent variable (response). This is best accomplished in an academic or research institute setting, where there is time and funding devoted to testing a research hypothesis. It may not be appropriate to complex systems where there are many independent dynamic variables operating simultaneously that confound the results. It is well known that even double-blind clinical trials of pharmaceuticals and medical devices may have delayed real-life adverse effects. Double blind means that neither the investigator (researcher) nor subject (patient) is aware of who receives an active agent and who receives a placebo. The main advantage of scientific research is that replication (duplication of the study) can provide assurance of its purported results, and that each study gradually advances the frontier of the scientific endeavor.

Empirical research

Perhaps the most productive method of human factors research has been the use of observational techniques. *Intrusive* observations may be made, where the subject is aware of the observations and may be influenced by the process, or *unobtrusive* observations may be used, where the subject does not know or suspect that he is being observed. Unobtrusive observations may be made by the use of one-way mirrors, peepholes, or hidden television cameras that may be high speed, high definition, and located far away. This may be called empirical research when the analysis of the observation is accompanied by special knowledge of the activities, an appreciation of the problems in conducting objective research, an absence of bias, and an understanding of privacy rights (ethics).

Intrusive observations may be desirable or needed. In a study conducted by one of the authors (Peters et al., 1962), the operational procedures for two complex items of maintenance equipment were evaluated on preproduction items. The step-by-step procedures were described in printed technical manuals. Some personal interaction between the research investigator and the subjects (trained maintenance mechanics) was necessary. On a hydraulic pumping unit there were 267 missing procedural steps, which accounted for 47.8% of all changes to the text. On a booster engine checkout and trouble analysis, changes to the text for clarity and deletion of extraneous words constituted 41% of the 307 test changes.

In an unintrusive observational study, the authors of this book carefully observed the drivers of 911 passenger vehicles in Ohio, California, and Ontario, Canada. They found that, in each location, about one third of the drivers had their head restraint (head rest) in a low position that would be unsafe in a rear-end collision. The head restraint reduces violent head motions and helps to keep the head and upper torso in alignment. The study was simple and direct, virtually uncontrolled, and performed without the drivers' knowledge or consent.

Interaction is desirable in evaluating troubleshooting and emergency procedures, whether they are in the form of technical manuals, checklists, audible messages, or computer assists. Similarly, there may be a better understanding of communication, information, and decision-making needs by observations and personal interactions. This is also true to verify accident reconstruction and customer difficulties.

In terms of situation awareness studies, there may be knowledgeable experts who prescribe what the operator should have known about his surrounds and presented cues in order to appropriately respond to the situation. However, the operator may have a very different perception of the situation that provides a different reality. What is important to him and needed for prompts may be discovered by observations and interactive interviews.

Descriptive measurements

Simply measuring human body dimensions, ranges of movement, and the kinematics has provided very useful information (Garret and Kennedy, 1971). Proper application of anthropometric data may provide a best fit in terms of foreseeable human dimensions, arm reach distances, hand force capability, and other human variables. In other words, it reduces the likelihood of "can't reach" situations, "can't see" over the dashboard or console, or "can't push" foot pedals with sufficient force. It may help prevent accidents caused by access doors so small that the eye cannot see what the hand is doing inside the equipment during servicing operations. An understanding of the principles used in epidemiology is helpful in collecting, analyzing, and interpreting descriptive measurements.

Task analysis

It may be advantageous to determine exactly what jobs and tasks are to be accomplished by equipment operators, maintenance workers, repair personnel, and others who may come into contact with a product, system, or process. At the beginning, during concept and early design, the analysis may be only a rough description of prospective tasks. This may be enough to plan what may be needed in terms of an effective human factors effort. During the equipment development process, more specific task information may be developed from component design specifications and drawings. During final development and availability of mock-ups or prototypes, the required tasks can be verified, elaborated, trimmed, simplified, or clarified.

Task analysis serves to identify and then break down a job or task into its ever smaller component parts. Patterns, sequences, and linkages may be discovered and analyzed. There are various formats used to list each task, who performs it, what decisions must be made, what type of action is required, the needed skills and training, the task criticality, its difficulty, and the need to coordinate with others. There may or may not be a determination of the likelihood of human errors and equipment malfunctions, but if so, it is usually restricted to qualitative estimations on a 10-point scale, with identification of the hazard as directed to people, the machine, or the intended purpose, mission, or required system performance.

Since task analysis can generate excessive paperwork and accounting-type workloads, it is often performed in a somewhat delayed and fragmented fashion, dealing only with possible critical, unusual, or dangerous tasks. For high-reliability projects, the assurance given by complete task analyses may result in a mandatory assignment of task analyses to human factors specialists, logistics analysts, or field support personnel. In any case, task analysis

can be a valuable asset whether it is formal or informal, systematic or fragmented, or performed only when operability problems arise. It also has
considerable value in helping to formulate detailed procedures, checklists,
cautions, and rules.

Simulation

The word *simulation* means to make believe, pretend, and simulate an
appearance without the reality. Aircraft cockpit training devices were used
to train pilots without experiencing real-world crashes. Human factors
experiments were then conducted with those trainers. Automobile driving
tasks were simulated by using a portion of a vehicle while a moving highway
was projected in front on a screen. These human research devices may have
been a gross simulation of the real thing, but they were useful at the time.
There are increasingly sophisticated simulations, used for research, that are
very realistic. A prototype vehicle or a mock-up of a product provides realistic opportunities for research and development. Extrapolation to the real
world is always a problem.

Instead of using costly hardware for simulation, there have been
attempts to use mathematical models of the human dimensional envelope,
human reaction times, and human decision making under various information and communication situations. Automobile accidents are routinely simulated in digital animated videotaped reenactments depicting the occupants'
biodynamics, rationalized human behavior, and documented injuries.
Graphics are a form of effective simulation. The use of predictive mathematical models is consistent with computer-aided design and computer-aided
manufacturing programs used to shorten design schedules (see Peters, 2002,
chap. 9, for additional information on human simulation).

The synthetic environment (simulation) may be low fidelity (lean) or
high fidelity (rich), interactive or responsive only, and centered on research
theory about decision making or, perhaps, on effective multinational military
deployment. It may be computer or Internet based, experimental or exploratory, have a good or poor pedigree, and be cognitive engineering or engineering psychology oriented. The cognitive studies may involve cueing,
spin, filtering, pinpointing, and situation awareness relative to friends and
foes in an information warfare setting. They may deal with purposeful
confounding, degrading, and denial of elements in the flow of information
as contrasted with robustness, protection, and dominance in terms of belief
systems. There have been studies of the logic of decision and the effects of
spoofing, deception, downplaying, hidden intentions, lowered capability
projections, and assaults on traditional values. This is illustrative of the
research potential of various forms of stimulation.

Focus groups

Human factors specialists may convene small groups of people to focus on
a particular product or topic. The people (respondents) are independent of

the company and usually a population of users or purchasers. They are selected to fit special categories (demographics) or subpopulations of interest. They may meet in a home (an informal setting) or in a specially equipped office conference room (a formal setting). They are usually paid a small sum for attendance. There are firms that can supply the people, but to avoid serial participants, the broad reach of the Internet is generally used to recruit inexperienced mainstream participants. In essence, they form a mock jury to evaluate customer reactions.

At first, the participants are encouraged to chat, become relaxed, and enjoy the social setting. Then more direct questions (interviews) ensue, and they may be asked to test products (observations) and give their spontaneous reactions (commentary). They may be provided snacks, pizza, and soft drinks during the 3- to 6-hour session. The human factors issues may relate to foreign markets and their cultural values, habits, and desires. They may involve advertising and marketing, including the accuracy and implications of translated language and pictorials. These user opinions may be quite different from what company employees are likely to say or what is obtained in operability (usability) laboratory settings at a factory site. Written surveys, based on the focus group commentaries and behavior, may serve to help estimate the extent of such opinions and their accuracy, and obtain information on narrowed subissues.

Design reviews

Human factors specialists generally participate in formal design review sessions conducted during the detailed design stage and before release for test or production of components, assemblies, and subsystems. This provides an opportunity to become better acquainted with the latest design version of equipment items, to present information on relevant human factors issues, and to determine what should be accomplished during the remaining scheduled time. The design review is attended by various design groups, subject discipline specialists, and managers. Action items are generally recorded and followed up. The human factors specialist should have his own follow-up in terms of contacting the responsible design groups and discussing possible human factors improvements to the design. Design reviews are early enough that there is less resistance, less cost, and less time required for design improvements.

Field studies

The human factors specialist should have a special interest in any postsale field analysis of products, processes, or systems. This marketplace experience may reflect the effect of misuse, abuse, unintended use, malfunction, or confusion from operating, servicing, and maintenance instruction. Customer dissatisfaction and human error problems may be discovered. The old "sell and forget," "buyer beware," and "due diligence" approach was quite different from the current "customer satisfaction" and "user expectations" concepts.

These new objectives require special field studies to complement warranty, adjustment, and field failure reporting systems. It is not uncommon for human factors specialists to conduct their own field studies to identify human factors problems and to determine how they may be corrected or prevented.

A field-type study that duplicates the use situations is often conducted by human factors specialists before production release. Consumer products may be tested in special homes and in cultural localities. In field studies of military land mines, inactivated (inert) mines were placed in the path of office workers and along trails in wooded areas. The objective was to determine general detectability by uninformed civilian personnel and the visual detection of various lengths of trip wires by informed military personnel. Automobile companies provide current model production vehicles to some employees in order to determine potential problems and customer satisfaction. Early field failures are best detected at an early stage by having close factory contact with the users of their products, rather than relying on distribution chain reporting that may be muted or ignored. Field studies conducted periodically during the product service life and thereafter can produce vital intelligence.

Meta-analysis

Meta-analysis is a retrospective method for combining and quantitatively synthesizing information from a variety of different research studies on the same topic (Peters, 1998). It may help to explain sources of bias, confounding, and heterogeneity in apparently conflicting or inconclusive studies. By combining data from small to moderate sample size studies, the enlarged sample size may permit the identification of small effects. Most meta-analysis studies have been based on experimental studies where biases have already been minimized by controlled randomized studies. But meta-analysis has also been used to combine nonexperimental observation data. A well-defined protocol is essential. The ultimate purpose is to obtain more meaningful results, but meta-analysis can also have prospective value in formulating somewhat similar research studies.

Stabilization research

Human factors specialists and industrial engineers frequently attempt to improve manufacturing process control. It may be as simple as providing a lift table for component parts at the assembly line, an articulating arm to move a heavy part onto an assembly, or a cord hanging from above to keep a tool adjacent to where it is used. At each work station there may be a difference in operator task performance variability. The unwanted human variance is often referred to as a noise factor that should be reduced and controlled (Peters and Peters, 2005). Stabilization means control of an unwanted variance or noise. There have been attempts to integrate statistical process control with designed experiments (Leitnaker and Cooper, 2005). This is a step ahead of the elemental use of control charts to reactively monitor process variation. It is an attempt to understand the process and causes of output characteristics that are critical for process improvement. When including a human factors experiment, an attempt should be made to ensure relative variable stability and good measurement procedures. The use of experiments in an industrial setting holds great promise with the advent of increasing computer power, sensor developments, predictive condition monitoring systems, and concepts of artificial neural networks as modeling techniques for complex functions.

Stabilization is also important in product design. There have been attempts to make the product insensitive to noise, whether ambient conditions, aging, wear out, imperfect materials, or customer use. This has been called robust design, which is a selection of design parameters to reduce product variability over the service life of the product under varying customer usage patterns. The objective is to stabilize the system.

Illustrative guidelines

Clutter

Clutter is the unnecessary information that can slow, confuse, or distract the information-gathering or decision-making process. It could produce an overload of information. The remedies include removing unneeded precise markings on a visual dial or gauge and the excessive words in an auditory message or warning. It could include deleting extraneous detail from instructional and procedural media. Replace it with only that which is demonstrably needed for context meaning and monitoring and to initiate action. An on–off signal light might reduce the interpretation that may be required with a precise and costly gauge. In general, simplicity is better than unnecessary complexity.

Visual references

Visual references are signals that provide spatial orientation as to location, direction, and distance. In darkness, without a visual reference, someone could go in the wrong direction or become lost. False sensations are common for aircraft pilots and have led to crashes. For example, when rapid acceleration feels to the pilot as if the aircraft is climbing and needs to level off, the corrections to the flight path could lead the aircraft into the ground. The remedy might be ground lights at night or appropriate cockpit instrumentation. Highways use lane markers, road edge lines, and other reference systems. In a restaurant, a dark staircase may lead to a fall if the edge of each step does not have a light strip, white border line, or phosphorescent reference marker. Luminescent (glow-in-the-dark) rung covers on the aerial ladders of fire trucks improve the visibility of the rungs. In general, blind movements are undesirable; a visual reference or guiding light can provide the necessary spatial and directional cues.

Fatigue

Human fatigue is a weariness, drowsiness, exhaustion, or worn-out feeling from physical labor or mental exertion. If an operator is fatigued, he may be less self-critical, more prone to false sensations, and subject to coarser motor activity, and may engage in rough mental approximations. If fatigue occurs during normal operations of a machine, the remedy may be in the design of a less demanding machine operation. A poor shiftwork schedule may have components that can be changed as a fatigue countermeasure. Since fatigued workers are to be expected, the design of equipment should accommodate that fact wherever possible. In general, good human performance suffers when a person is tired or fatigued.

Accidental activation

Accidental activation of a control device is the inadvertent and unwanted movement of a control that serves to turn on, cause movement, or otherwise actuate a machine. An operator's hand or arm may touch or brush a control knob, lever, push button, rocker button, slide switch, pointer knob, thumb wheel, crank, joystick, or toggle switch with enough force to move the control. A blouse or shirt sleeve may catch on the control. Someone may lean against or sit on the surface containing the control.

The act may be intentional, but it could be a misdirected choice between controls (hitting the wrong button). The remedy is to use a cover, recess, skirt, or isolated location. It may be appropriate to have a key lock, interlock, or control event sequencing. The key should not be removable except in the off position. The control may be given increased resistance to activation and be shape coded to reduce the likelihood of mistaken activation. Accidental actuation of a critical control is less a user problem than a design problem.

Control motion compatibility

The direction of motion of a control, such as a steering wheel, hand wheel, or lever, should be the same or directly related to the expected response movement of a vehicle, display pointer, or the physical object being manipulated or controlled. A wrong response movement may cause confusion and promote a lack of eye–hand coordination. The remedy is to use standardized or stereotypical responses to control movement and to ensure compatibility within a workstation. If a control lever is moved upward or forward, the responsive function is to turn on, raise objects, extend, or increase motion. A clockwise movement of a knob, rotary switch, or hand wheel is to close valves, cause movement to the right, or increase intensity of the response. A locking key should be in the off position when vertically oriented. In general, the design should follow user expectations, since it is those expectations that become manifest during actual usage and in emergency situations.

Operator incapacity

Operators do have heart attacks, loss of consciousness, or even death at the work site. If there is an incapacity or inability to control what is happening, serious accidents can and have occurred. The use of dead-man switches on railroad trains means that when the operator falls away from the dead-man control or does not exert sufficient force, the train automatically brakes to a stop. Another device is the railroad train vigilance button, which requires the operator to press the button and reduce speed when passing a slow or stop sign. If not pressed, the train will be automatically braked. In general, there should be an automatic system shutdown of critical conditions if there is any operator incapacity.

Abuse

The engineer who helps to design a product may believe the reckless, careless, improper, and potentially damaging use of his product is blameworthy reprehensible abuse. The customer may have a very different viewpoint, and his beliefs and expectations are paramount in a customer-driven marketplace. Since it is predictable that abuse will occur, the human factors design guideline objective is to have products that can withstand a reasonable amount of abuse during shipping, handling, installation, use, relocation, updating, alteration, repair, recycling, dismemberment, and ultimate disposal. Fragility is not acceptable if a damage-tolerant or abuse-resistant product is technically and economically feasible.

Coding

Controls and displays should be clearly coded for identification purposes and to discriminate one from another by color, shape, size, position, feel, or labels. Color coding may provide a specific meaning, such as red for stop or a red mushroom button for emergency use. Shape coding may provide visual (sight), tactual (touch), and representational (looks like) identification. Position coding (location) separates groups of controls by function and provides spatial cues. Size coding provides gross tactual cues. The surface of knobs can be smooth, rough, serrated, or knurled for a different feel. Coding is particularly important for multitasking equipment where there are many optional and add-on devices and mistakes are easy to make. Fire trucks have color-coded pump controls where there is more than one pump and hose. Aircraft use standardized control shapes so the pilot can maintain a look-and-avoid vision out the cockpit windshield. Aircraft stick shakers are used as stall warning devices. Valves may have a definite feel or click into the correct position. In essence, coding and cueing are a vital part of the human information system since they help to prevent mistakes.

Pictorials

Pictorials are miniature representations of objects by the use of pictures, electronic displays, or mechanical (physical) objects. They need little interpretation and can quickly depict spatial relationships. Generally, the operator's display is fixed in position and the others can be moved or manipulated. A graphic representation of a complex array of parts, instrumentation, displays, and controls on a large display board may convey status information quickly and accurately, can show direction of flow, and may be animated electronically.

Grouping

Controls and displays necessary to support an activity or function should be grouped together and arranged by sequence, frequency, and importance. Critical displays should be in the primary visual zone, highlighted, consistent in application within a system, and located at an appropriate viewing distance.

Accessibility

That which is needed at an appropriate time should be conveniently located for human use. In fire trucks, the packmule (hosebed) may be lowered to an ergonomically appropriate waist level so that firefighters do not have to climb onto the truck to unload the hose. A hydraulic system can help to reroll the hose without muscle strain.

False cues

The cues that an operator relies upon may be false and could lead to an accident. Rapid aircraft acceleration causes the pilot's vestibular otolithic hair cell filaments to move, resulting in the sensation of climb. When the pilot tries to correct and assume level flight, he actually dives and goes downward toward the ground. Other false signals are from rapid head movement during descents and turns, false vertical illusions after rolling or banking an aircraft, and disorientation when flying over sloping cloud decks or mountain slopes. Flight instruments and pilot training are designed with false cues in mind. However, false cues and illusions are fairly common in the use of a variety of products, and they should feature a design antidote.

Abstract theory

It has been found (Peters et al., 1962) that many training courses for technicians and service personnel have a high content of abstract theory and material that is quite remote from that needed. The emphasis should be on how-to-do-it training involving physical contact with the actual equipment. More than one person in a training group should be used for demonstrations with the physical equipment, rather than most of the trainees relegated to the role of passive observers. The training should involve functional relationships between equipment items, the causes of trouble, and maintenance skills. The training should be preplanned, systematic, and conscientiously supervised. Subsequently, good training programs could be videotaped for future use and commentary.

Ignoring documents

It is well known that people will ignore or, at best, quickly flip through instruction booklets, service manuals, and technical publications. They usually try to assemble or use something by trial and error, and only when defeated will they look at the printed instructions. Mechanics quickly commit procedures to memory and then ask other mechanics for help when needed. The reluctance to use technical publications is only human, and the design engineer should not rely upon compliance with complicated and complex written instructions or computer printouts. The design of a product or system should build upon the human's intrinsic ability to follow simple, logical, step-by-step procedures that are made obvious. Improper procedures should be anticipated and thwarted by design. Use of inexpensive modern visual and auditory media could be used where procedural or instructional needs remain after design. Technical documents may be prepared for reference purposes, as a teaching aid, as a job refresher, as a field checklist, or as a job-sequenced manual with functional, print, or part number call-outs. Technical documents should be given a technical review for clarity, accuracy, logic, school grade level, sequencing, completeness, methods of updating, comprehension by customers or users, and validation as to their purpose.

Information

The information needs of an operator of a machine should be clearly defined at an early stage of design so that appropriate changes can be made before the design is frozen. If the operator tasks are left to the prototype stage, the operator may be tasked in such a way that there is overload, confusion, and an inability to discriminate what information is really needed to appropriately initiate action in a complex systems context.

Cold weather

When equipment is to be used in cold weather, it should be designed to accommodate stresses created in both machines and humans. Assume that the human body has limited heat reserves and heat production capability. Maintenance actions, emergency repairs, and normal operations should be capable of being completed within a short defined time interval. In other primitive conditions, the object may be to avoid human heat stress.

Audible alarms

Audible alarms serve to warn people of impending danger, to alert operators of significant malfunctions in a system, to remind or cue the operator to critical changes that must be made in a system, and to supplement visual alarms (as in automotive vehicle air bag warnings). They are used as fire alarms, for notification of possible exposure to toxic agents, and to alert a ship crew of an impending collision. The alarm may signal by a horn, buzzer, siren, bell, whistle, or beeper. The frequency range may be between 200 and 5000 Hz, with lower frequencies used for longer distances, indirect pathways, or passage through partitions. There are situations where there are multiple alarms on forklift trucks in industrial storage and fabrication areas. In such situations, clearly discernible differences between alarms should be made by spectral (waveform) composition, temporal (time) duration, intensity, or sound patterns. A sense of urgency can be communicated by the coding selected.

Intelligibility

The intelligibility of speech refers to its effectiveness in communicating messages in the context of unwanted sounds or noise. Noise also hinders mental concentration, and those exposed become irritated, inhibited, nervous, and fatigued. To improve intelligibility, speech transmission equipment should operate in the speech spectrum of 200 to 600 Hz (as contrasted to the spectrum of 100 to 8000 cycles per second for normal conversation). The dynamic range should have signal input variations of at least 50 dB. Improvement may be obtained with noise canceling microphones, noise shields, filtering, squelch controls, and binaural headsets where the background noise is above 85 dBA.

Installation

Replacement of an electronic component or subassembly should be facilitated by installation instructions. Electronic components should be marked to show the direction of current flow, so the component will not be installed backwards (reversed). This rule also applies to subassemblies that should be designed in a fashion so that there is only one correct location possible for installation. Electrical sockets should have pin (prong) arrangements or keyed contact surfaces that prevent incorrect connection and installation.

Arming

The electrical arming (activation) of an ordnance (explosive) circuit should have a physical blocking device to prevent inadvertent arming. An out-of-position indicator should be used to signal the armed position and readiness to fire (actuation). The blocking device and position indicator should be readily observable by nearby personnel.

Balance

Large containers should be marked with the load (center of gravity) centers, because workers cannot see inside the container. A forklift truck that picks up an off-balanced load with its prongs (forks) could easily tip over. A material transfer device could be damaged by an inbalanced overload. The weight, balance, and need for tie-down straps during transportation should be indicated conspicuously on transportable containers.

Overload

Machines should be designed to prevent operation with overloads or during excessive speeds, reach distances, or production. In addition, there should be placards indicating safe load capacities (load limit charts), proper operating speeds, and how to obtain relevant operating manuals. Reliance on add-on safety devices (safeguards) is ill-advised, particularly when the device is manufactured by others. Crane upsets (overturning) often occur because the operator does not extend the outriggers to stabilize the crane. Some cranes now have automatic hydraulic system extenders with interlocks to prevent operation without the outriggers being extended into proper position. There are proximity warning devices for use when the crane operator moves the boom and it intrudes into the danger zone of high-voltage electrical power lines. The overloads, outrigger problem, and power line contact situation are all common foreseeable human error situations. Overloads may combine with other factors to produce system failure.

Controllability

The human operator should be able to control the machine he operates from start to finish with no loss of control. That did not occur with some SUV light truck vehicles that drive differently from conventional passenger vehicles. They may have factory-installed placards indicating an increased likelihood of rollover. In fact, the vehicles slide out (the rear end rotates sideways as the vehicle oversteers) and the vehicle may be tripped into a rollover if the right rear tire blows out or experiences a tread separation. The reason it goes out of control is that the movements of the vehicle are faster than the corrective reaction times of the driver. The vehicle lacks controllability because of the designed-in vehicle dynamics resulting from a high center of gravity, narrow body width, and short vehicle length (between axles). The tendency to tip to one side and oversteer could be improved by torque rods that shift downward forces to the opposite lifting side, a stiffer suspension resistant to tipping, and an independent suspension. Where a lack of controllability persists, the vehicle could be stabilized by the use of an electronic stability system that exerts braking forces and other inputs faster than the human capability. Other automobile features that compensate for human limitations are blind-spot detection sensors that warn the driver that something is in the blind spot near the C-pillar. There are lane departure warning devices that tell the driver that the vehicle is shifting out of the lane by vibrating the seat and the use of chimes or indicator lights. There are adaptive cruise control devices that will automatically slow the vehicle down and shut off below a set speed. There are collision mitigation braking devices that pulse the brakes as a warning before a crash and automatically close the sunroof and windows and tighten (retract) the seat belts. All such devices compensate for limitations in human capability.

Unusual conditions

General (small) aircraft pilots are familiar with unusual conditions that could cause an out-of-control situation. The pilot may fly into severe turbulence, may be distracted or inattentive to his cockpit duties, may experience spatial disorientation, or may suffer from instrument malfunction or failure. The unexpected often occurs in the operation of various machines and factory processes. If something unusual occurs, there should be discovery (catch it) and recovery (corrective) procedures to forestall entry into a critical situation that could lead to serious injury and damage. The questions for the design engineer are: What would happen if …?, How could the product or system be saved?, and What should the operator do? Unusual conditions are not as rare as many people believe.

Prompts

Prompts started as cues to stage actors for words, topics, or actions that might have been forgotten. Public speakers may be prompted by headphones, written notes, teleprompters, or other reminders. Equipment operators may see or hear prompts intended to command their attention and encourage action. A flashing light or a beeper may prompt or warn about an out-of-tolerance event on a graphic board representing the process flow in a factory setting. A troubleshooting device may have prompts to signal the search direction in which there might be possible faults. A digital camera may have prompts to indicate the steps necessary for a proper exposure. Thus, prompts attract attention, serve as reminders, and can help guide someone in step-by-step procedures. The human brain, through the hippocampal and related structures, mediates context-dependent learning and episodic memory (Hargreaves et al., 2005). The neural circuitry is subject to changes in intercell synaptic neurotransmitters, cell surface physical plasticity, intracell vesicle recruitment, system circuit enhancement, and cellular morphogenic changes (Peters and Peters, 2002a). This dynamic process is accompanied by short- and long-term memory changes. Therefore, it is understandable that memory prompts, episodic or sequential, are needed by human operators of complex systems.

Self-governance

Similarities

There are some important similarities between the work of practitioners of human factors engineering and those of human error reduction. Each are specialists by virtue of education, experience, skill, talent, special unique knowledge, and focus. They may act as technicians doing work with no independent discretion, or they may act as self-energizing professionals. They both may be requested or assigned to study a problem by others. Professionals also act by understanding the overall objectives and determining what can be done in terms of what they deem necessary to further the known objectives. Technicians may deal only with the obvious, but professionals may deal with all probable alternatives. Technicians may keep to themselves, but professionals may coordinate with other specialties where reasonable and appropriate. Self-governance is the most desirable and valuable approach resulting from professionalism.

Overlap

The following case example serves to illustrate the overlap between human factors and human error:

> At an offshore gas platform in Europe, on June 24, 2000, a contract worker (a production technician) placed his foot on the handle of an isolation valve. The force was enough to open the unsecured valve and result in an uncontrolled release of hydrocarbons. The worker was struck in the face by the stream of gas and suffered permanent injuries. An investigation indicated the cause was human error. An interlock on the valve had been removed from the isolation valve while the worker was refitting and calibrating a pressure safety valve. A judge in England indicated that the company's risk assessment, required by regulation, did not consider the valves (SHP, 2003).

Scope of work

Since the scope of work for the error specialist is usually quite broad and often crosses the boundaries of an organizational structure, the overt support for such activities should be in either a documented plan of action (a program plan authorizing the activities) or by working agreements reached with organizational elements that might be involved. In the process of achieving a mutually agreeable work authorization, it should become obvious whether managers wish to exert loose or tight control, what the expectations are concerning communications where local approval of changes does not occur, and what might be required in terms of diplomacy. Professional behavior is desirable when working with others, and this includes basic civility, respect for others, confidentiality, absence of bias, creating a circle of trust, providing beneficial results to others, and manifesting loyalty to the corporate enterprise, the integrating contractor, the funding agency, or the research institute. In essence, civility reflects the basic character of a discipline. As time passes, an activity gap analysis may be needed to assess progress and to determine the gaps that may exist and what should be done about them (such as an updated plan). The plan should ensure some measure of self-governance.

Moral obligations

When dealing with design safety, those engaged in the human factors and the human error activities assume certain moral obligations as to the safety and well-being of others. It would be wise to determine the extent of such duties and the considerations that necessarily follow. The persons affected may be within the company, in the distribution network, at the sales and service organization, the ultimate customers and bystanders, and those who repair, upgrade, or demolish a product, process, or system. The question may be: What is the foreseeability of harm to each, the benefits and trade-offs, and the burdens and balancing that will occur as a result of the human factors and error reduction efforts? In other words, the moral obligations are presumed to flow to all others who may be harmed. These obligations can be best served by self-governing responsible conduct on the part of a professional.

Language interpretations

It should be understood that words and concepts used in one discipline may have a different meaning to those outside that sphere of reality. An aircraft pilot understands what a cockpit means to him, not what a gambler means by a cockpit fight outcome. The human factors engineer deals with average persons in the context of a normal distribution, Gaussian curve, or bell-shaped distribution frequency, and the average may or may not be defined as the population around the central tendency of ±1 standard deviation or 68.26% of the general population. The statistician may have three averages: the arithmetic mean, the median, and the mode. There may be a standard error of the mean (theoretical fluctuation of the mean). The reliability engineer may deal with different frequency curves and averages. The forensic specialist may deal with the average reasonably prudent person, which assumes a flat curve of ambiguous proportions and included population. Thus, does average include 1, 68, or 95% of the general population? Is error a statistical concept, an experimental concept, a descriptive term, or an expression inferring possible cause and cure?

There are differences in the meaning of words used by other disciplines, other conceptual realities, culturally shaded foreign languages, and marketplaces. These different meanings are very important for human error reduction in communications systems. A self-governance approach to the management of human error could help dispel maladaptive complications that could ensue from language interpretations, hidden meanings, myths, or misunderstandings. It could promote more relevant error prevention remedies.

Negotiating change

Design changes do not automatically follow the recommendations of a human factors or error reduction specialist. They must convince others of the merit of their proposed product improvement. This requires some negotiation (Peters and Peters, 2004). The duration, intensity, and content of this active phase is essentially self-motivated and is aided by a sense of self-governance. Some resistance to change is to be expected, based on the need for the rewiring of established neural circuits and intellectual comparisons to the rule sets of a globalized world.

Caution: In the past, a few specialists have converted self-governance into a license for arrogant behavior, demeaning acts to others, and extreme beliefs of self-importance. This is contrary to the required teamwork and good interpersonal relationships essential for negotiated product improvement.

Caveats

Constant change

By necessity, most product design efforts are a continuing iterative process. This step-by-step process means there will be constant change. A strong manufacturing quality objective is continuous product improvement or never-ending change. This suggests that the prevention of the causes of human error must also be a continual process, from product concept inception to final disposal.

Symbiosis

The contents of this chapter suggest that the prevention of human error should be a symbiotic process between various occupational disciplines and company functions. This is particularly true for human factors and system safety where enrichment and collateral benefits can occur.

Noise control

In terms of human error control, all signal–response relationships should be made insensitive to noise variations. Distractibility and signal identification should be tested.

Dynamic systems

In terms of human error, dynamic multiresponse problems may require a first-approximation resolution that could be demonstrated to be effective under real-life conditions by proper testing.

Multiple causation

In terms of human error reduction, where there is a plurality of causes for a human error, Occam's razor suggests that multiple causation should not be initally postulated unless absolutely necessary. This suggests that the removal of one major cause might eliminate the human error occurrence, that a more precise identification of error would narrow causation, or that factor analysis testing is appropriate.

Uncertainty

Human error problems may be resolved even with uncertainty, based on incomplete information, but only if verification testing is scheduled to confirm or refute the hypotheses involved.

Process shifts

Human error as an assignable cause of a change or shift in manufacturing process control requires the designation of a monitoring window to determine out-of-control signals, their cause, and verification. Controlled experiments are possible during process control.

Field reports

Human error detection from field tracking studies may be costly and untimely, so earlier screening and testing are preferable. Design, procedural, and technical communication modifications are almost always easier and less costly early in the design cycle. Some life cycle auditing is still necessary.

Bias recognition

Caution should be exercised and bias recognized for certain techniques dealing with uncertainties. Meta-analysis may be biased if based on published studies, since journals favor positive results. Value-of-information (VOI) studies substitute expected benefits for uncertainties yet to be resolved. If decisions must be made based on uncertain variables, this fact should be clearly stated.

chapter eight

Testing and functional validation

Introduction

Even the best analytical methods are a shade removed from the real world of anticipated equipment function and predicted customer usage. At each step in the product or system design and development process, as real hardware and software become available, verification tests are performed to determine what works and what does not. What does not work to expectations can then be changed and improved. Prototypes can be patched or redesigned to meet specification criteria, customer demands, and company policy requirements. Design is fundamentally an iterative process with each step checked against real test results. Testing tends to remove uncertainty concerning product performance.

The most meaningful test occurs after the product or system has left the factory and is being used by the purchaser or subsequent owners. These tests serve to validate the function of the product or system in terms of the aftermarket objectives, customer satisfaction, and from the perspective of future purchasers. Unfortunately, once a product is sold, the buyer beware philosophical bias has misled many manufacturers into a failure to systematically test to determine how well their products really work and if they really meet customer needs. In fact, there are some situations in which the only true test of product performance is the real-world functional validation, whether by astronauts, submariners, explorers at the North Pole, users in underdeveloped tropical geographical areas, antiterrorists responding to attack, medical specialists during an emergency health crisis, or those engaged in military combat.

Testing is often viewed as a pass–fail procedure. It is whether or not the test results meet or exceed some specified test requirements. What is actually tested may be quite limited in scope and in terms of what is measured and evaluated. The data gathering may be on issues of interest to the test engineer, but marginally related to the customer needs. The conclusionary opinions may have little merit in terms of error misuse, abuse, mistakes, or other errors. The error specialist may be able to add unique test requirements or define observational opportunities so that information can be gathered that is pertinent to error causation and prevention. This can be done on a nondestructive and economic basis. This chapter describes many of the tests that might become available and useful to the error prevention specialist. It also includes some basic principles relating to error testing that might be helpful in developing test plans.

General principles

There are some general guiding principles that should influence how human specialists test or participate in testing (as reported by Peters, 1963).

Direct observations

"Direct observation of actual working conditions is critically important to determine the true nature of the situation under study, to locate significant unreported or unrecognized problems, and to determine what could be an effective remedy. Analysis, theory, logic, and hypothetical assumptions are no substitute for direct observation."

Firsthand investigation

"You can trust your knowledge of human-initiated failure to the degree to which you are able to make a firsthand investigation of all of the facts in the case. A corollary principle to this is that upon investigation a human error 'detective' will usually see a problem quite differently than it is originally reported because his perspective (or what he is trained to look for) is different."

General solutions

"Beware of general solutions to classes of problems. Dramatic solutions to big problems are rare and usually misleading. There is much more assurance of constructive action when specific problems are analyzed in a specific fashion and when sufficient effort is budgeted for follow-through to ensure the attainment of a satisfactory remedy."

Definitions

The following terms are used to describe certain aspects of testing. *Verification* means that the test accurately measures that which the test purports to directly measure. The test criterion may be an immediate and arbitrary objective, such as a specified test result on a component. *Validation* assesses the test and its relationship to some ultimate objective, usually its real-life utility. It is a measure of how the product will meet customer needs. *Sensitivity* is the power of the measurement system, that is, its ability to discriminate and measure a desired variable. The measurements should be reasonably precise, free from extraneous confounding, and capable of useful interpretations. *Relevancy* is the relationship to a known, defined, external, and practical reference criterion. It relates to a useful external objective. Information is relevant if it has, by reasonable inference, a tendency to prove a fact — it is germane or appropriate to the situation or objective.

Illustrative tests

Proof-of-concept tests

During the preliminary or concept stage of a product or system, there should be a rough assessment of what the humans in the system may be required to do. It is easy to relegate consideration of the human interactions to a later date, because the early focus is usually on mechanical design, materials, manufacturing capability, and economic considerations. That is, is the concept technically and economically feasible? In the process, there may be neglect in considering whether the initial design is appropriate in terms of human performance, safety, and error potential. For example, the authors received a telephone call from an engineer during his break in the final design review of a complex and very expensive system. The questions related to fundamental issues that should have been resolved years earlier, rather than at the brink of its release to production. Some systems, such as passenger automobiles, use the driver's anthropometric and biodynamic dimensions as a starting point for interior design, which in turn affects the exterior styling and function.

Early awareness of possible problems permits calm and considerate evaluations of available options. There could be testing on older models of the product, on rough mockups, or by the use of computer simulations. The test objective is to prove or disprove the error potential and its consequences.

There may be a possible situation awareness problem, defined as a dynamic understanding or perception of the current environment in relation to relevant equipment parameters. This, of course, would influence an operator's decision-making or choice behavior (including action errors). If early in the design process, appropriate research data may enable significant improvement in the man–machine system. One study evaluated situation awareness measurement techniques, such as self-rating by aircraft pilots, observer ratings by independent trained observers, written questionnaires, online questions during simulated tasks or posttask, and high-fidelity simulation experimentation (Endsley, 1995). Situation awareness in a complex person–machine system may involve processing many simultaneous and disjointed tasks, although people can thoughtfully consider only one thing at a time, and a division of attention may result in a loss of sensitivity with increased bias and errors (Adams et al., 1995). Such studies, already published, may be of particular value during the proof-of-concept phase of product and system design. They identify usable research and suggest needed research or testing.

Component and assembly testing

The first completed piece of physical hardware may appear at the time of component testing. This could be the first opportunity to obtain test results, useful for error prevention, on actual hardware. The component may be designed and fabricated by the company or supplied by a vendor. The test requirements may be minimal or extensive, relevant to the component only or to the next higher assembly, and conducted by the company or by a first-tier supplier. Without the intervention and coordination of the error specialist, there might be little thought about possible error problems.

Typical errors during component testing include improper adjustment of mechanical linkages, wrong size orifices installed, valves not closed properly, improper installation of O-rings, hoses improperly connected, pins bent on connectors, bolts not tightened, and gauges incorrectly read. At a test station, a mechanic was assigned to monitor a circular oscillographic recorder, and when the pen crossed a red line, he was to depress the test cutoff button. The button was depressed inappropriately. It was found that chart watcher error was common at that test station but unreported.

At about the time that component testing begins, there are generally discussions and decisions being made about subsequent testing when groups of parts, components, and assemblies become available and functional. This includes statistical test design to minimize the overall costs of testing. This is the moment at which error reduction observations and test needs should be communicated to those planning product testing.

Parameter analysis

There may be critical work tasks that justify a parameter analysis. The parameter to be tested may be a single task, such as the torque (a tightening force) manually applied to a bolt on a pressurized fluid duct. Too little force could result in leakage and too much force could result in metal cracking during future use. Testing may involve the nominal or recommended torque that is desirable, the acceptable excursion limits (plus and minus the nominal), and the range of forces found to be applied by mechanics (test subjects). The forces applied beyond the limits indicate a human torque error and suggest that changes should be made to the job aids (torque wrenches) or to the bolt head size, shape, and strength. Where there are multiple parameters at play and error reduction is needed, it may be appropriate to use design-of-experiment techniques, using centroid and excursion limits, in order to optimize performance requirements.

Systems testing

Systems testing is performed on an assemblage of individual pieces of equipment and support devices that are intended to perform a given function, mission, or objective. The assemblage is treated as a whole, a system of interconnecting equipment, with people, and in a specified environmental surround. Systems testing measures, determines, or approximates system effectiveness.

The individual items may work as intended, but when connected and interacting, there may be malfunctions or failures of the system. The focus is generally on the interconnections and the human operators. The testing may be at the assembler's or manufacturer's factory site or as installed at the customer's location.

There may be separate operability, usability, or maintenance laboratories or facilities for more intensive or controlled studies. The test protocols, issues, subjects, constraints, and objectives vary widely.

At a *maintenance demonstration facility* the following types of problems were found:

- There was no coding or identification on wire harnesses, connector plugs, and terminals on an electrical unit. Maintenance requirements included disassembly and reassembly. Incorrect connections were likely, which would be called assembly errors. Murphy's law states that if anything can go wrong, it will.
- There was no "power on" indicating light for a hydraulic pumping unit. This created a situation in which unexpected things could happen to an inexperienced maintenance technician. The energized system could suddenly start from inadvertent actuation of a push button, the technicians could suffer electrical shocks when touching electrical terminals, and the errors might seem to be those of the technicians for careless behavior.
- A high-pressure pump and motor created an excessively high noise level. This annoyed operator personnel, creating stress. Voice communications were difficult. Errors were predictable under conditions of stress and possible misinterpretations of communications. The motor was rewired, insulation blocks were installed, and flexible lines were installed to reduce vibration transmission to other components.

In general, it was found that systems testing that utilized available factory technicians produced results far different than the results obtained using test subjects from a pool of actual consumers or ultimate users. In terms of training customer personnel, it was concluded that teaching abstract theories and general knowledge might not significantly help technicians who need to know the functional relationships between items of equipment, learn how to work as part of a team, and acquire maintenance skills rapidly.

A maintenance demonstration facility may have its own design-for-maintainability checklist and be gathering data for estimates of mean time to repair. There may be other maintenance-oriented objectives, such as the development of preventive, scheduled, and unscheduled work procedures, with skill levels, time required, and identification of work that demands good judgment. The conceptual approach may be similar to evaluating a pilot run on a process that has to be tuned to accommodate many variables and to achieve optimum properties and performance capabilities.

Special testing

There may be special tests, unique to the product, company, or industry. These tests may take place when either prototypes or production-type products or systems become available. For example, in the automotive industry, there may be vehicle crash testing to determine crashworthiness, rollover testing for tip-over propensity, component sled testing for force resistance, proving ground testing for accelerated stress, and overturning tests for agricultural vehicles (stability on slopes). The aircraft industry may conduct passenger evacuation tests, oceangoing ships have shakedown cruises, and pharmaceutical products are evaluated in three-stage clinical trials. Each provides an opportunity for error reduction in the product and in the test methodology.

Intentional exposure to harm

There have been test situations in which humans have been knowingly exposed to danger in disregard of known ethical principles (see Chapter 12). An example is the periodic argument about the intentional dosing of human test subjects with toxic chemicals with or without their knowledge or full understanding (Hileman, 2005). The argument is that there should be human exposure data upon which to base regulations, standards, and safe threshold limits or machine operation. Participants in chemical trials have been orally administered neurotoxic insecticides, such as dichlorvos and aldicarb in the 1990s. There were human dosage tests in 1998 involving an organophosphate insecticide (azinphos-methyl) and roach spray (chlorpyrifos) administered to college students who were paid $460 each to swallow the pills. In 2004, there were tests with minority groups and college students using the possible neurotoxin chloropicrin, which previously had been used as a chemical warfare agent in the First World War. The argument is that in the absence of human data, the animal data must be extrapolated from a no-observed-effect level (NOEL), using a factor of 10 for the fact that humans may be more sensitive to the toxic agent than animals, another factor of 10 to account for sensitivity variations in humans, and a third factor of 10 for the increased sensitivity of infants and children. The total safety factor would be 1000. But human testing has been accomplished well below that safety factor and submitted as evidence of safe human exposure, with claims that a safety factor of less than 10 with humans is appropriate. Human trials (human test experiments) with small sample sizes may not yield statistically meaningful results or permit reasonable inferences. Ethical issues are becoming more important.

The basic issue is whether the endangerment of humans in tests should be permissible if there are no other means to obtain the needed information, or is human experimentation unethical if irreversible harm could occur to the test subjects.

Compliance testing

A product may be tested to determine if it meets, exceeds, or fails a company specification, an industry trade standard, or a government-imposed requirement. The testing may be accomplished by the manufacturer, independent laboratories, universities, insurance-funded entities, private parties, and suppliers. Compliance testing is usually on narrow, carefully defined issues in which a consensus between all interested parties has been achieved. There is often considerable potential for data gathering on other issues. The information in test documentation may be compared with other test material and relevant engineering design drawings. Errors may be the subject of informal complaints, but are rarely reported.

Installation testing

Complex systems, in their entirety, should be tested prior to shipment to a distant location for installation and use. This may not be possible where there are several manufacturers of subsystems and with an integrating contractor at the location of use. In this situation, the final system testing is accomplished at the operational site using various company personnel, customer personnel, or a mixture of both.

The first type of testing occurs during installation. For example, a large plastic injection-molding machine, with very large dies, was transported to the customer site in pieces. It was assembled by manufacturer technicians and service representatives. There were trial runs over several months, but the customer refused to accept the machine because of the uncontrollable variations in the plastic product being produced and the difficulties in maintaining production on a 24-hour, 7-days-a-week schedule. The claim was unreliability of the machine and undue demands on the operators. The company technician could operate the machine, but nobody else could.

Another example, in the same industry, was a plastic injection-molding machine that was designed in one location and assembled at a customer location as the pieces were manufactured by suppliers and delivered to the customer site. After numerous attempts to assemble a working machine that could meet the promised capability, the machine manufacturer abandoned the project. During the design and assembly of the machine, the learning process resulted in many errors and attempted corrections, but the overall compounding effect of the errors simply overwhelmed the manufacturer.

In both cases, there were major error problems that overcame the benefits of a new design. Error prevention was beyond the knowledge and capability of the machine designers.

Error testing should be accomplished during the on-site assembly and the trial production runs of a complex machine. Such testing also should be accomplished during the initial installation of any completed complex machine, and when possible, performance problems could be anticipated. The error specialist may be able to render what is considered an unacceptable machine, because of difficult-to-operate problems, more acceptable in terms of customer satisfaction.

Field audit and validation testing

After a product has been sold, delivered, and accepted by a customer, there may be aftermarket questions, such as: How good was the product?, Were there any troubles with the product?, Did it work as intended?, Could it be better?, Was the customer satisfied?, and Would the customer buy another product from the company in the future? Some companies remain passive, waiting for indirect answers from customer complaints, adjustment data, warranty feedback, or reviews in trade publications. Adverse information may be received too late, which is late enough for the company to be on the defensive.

Other companies have some of their employees use the product and report back to management. Some use focus groups composed primarily of current users of the product. Companies with field service technicians, in contact with the user base, may provide useful information. Product audits by independent research firms may have checkers or reporters go out and check for trends among the supply chain (vendors, distributors, exporters, middlemen, buyers, and resellers). Such information may be used to supplement other sources of aftermarket information. A more proactive, early, deliberate, controlled, and representative user population may be used by company-employed error detectives who visit and have direct customer contact. Even a small sample size, with direct and personal contact, may be revealing and enable changes to be made in a timely fashion. The direct interrogation of customers is needed as a functional validation test; all else may be remote, indirect, and fragmentary.

Modification testing

Many products and systems are modified, improved, or altered during their use. When the product becomes different, it should be tested. Foreseeable changes may exist by virtue of popular aftermarket modification kits, silent recalls, better replacement parts, or intentional customer changes to meet their unique needs.

An overhead bridge crane was to be installed in a prefabricated steel building. A new structural frame for the crane was needed. Two contractors submitted bids. The least expensive bid was to bolt the structural frame to the floor using base plates. The second bid was to cut the floor and install a concrete footer for each structural column. Normally such buildings have a 4-inch reinforced concrete floor, but it can vary from 1 to 6 inches in thickness (Dewey, 2005). To prevent a mistake and possible disintegration of the floor from a base plate overload, the as-built drawings and specifications should have been reviewed to determine the actual floor thickness or a core drilling of the floor performed where the structural columns were to be placed. Modifications to buildings are common over the course of many years, and the actual final construction may be quite different than that indicated in the original architectural drawings. Errors can be prevented by looking at the as-built drawings to determine what may be hidden from view.

If there has been integrated steel design, which is a computer analysis and three-dimensional modeling of the building structure, more informed, precise, and accurate decision making is possible. Merger of structural drawings and erection drawings into a single document also can help. The miscellaneous steel, such as handrails, stairs, and ladders, may be included. The computer modeling permits the erector to coordinate with the engineers and includes error reduction features such as reach holes for access to tight areas, sufficient nonshared bolt connections, and appropriate lifting lugs for modules. In essence, testing for building modifications should be avoided, because of cost, when as-built drawings and updated computer analyses could provide the same information.

Special needs testing

There may be a special need for verification or validation test information in response to customer complaints, threatened recalls, or unusual warranty claims. It may be in response to government regulatory or industry standards-making deliberations. It may relate to countries that mandate product updating, recycling, parts reuse, or destruction. It might relate to accident reconstruction and injury causation. There may be a new use of a product, akin to off-label and experimental pharmaceutical uses. Human error is often a central feature of such special needs testing.

In the driveway of his residence, a man jacked up his car in order to drain and replace the motor oil. When he was under the vehicle, it slipped off the jack, crushed his chest, and he died from the injury. The bottle jack had a very small base and was unstable if a load moved slightly. The safety problem with the jack was well known to some people, but the jack manufacturer did no special testing to confirm or deny the problem. The jacks were not recalled. The manufacturer was notified of this death nearly a year after the accident. During the long delay, there were numerous incidents that could have been avoided by special testing, recall, or modification of the product to have a more stable base.

A change in the state insurance code mandated that claims adjusters, in automobile accidents, not reveal the address and telephone number of their insured to others involved in the accident. This was to prevent private discussions and inadvertent admissions out of the presence of legal and insurance representatives. It was also intended to prevent physical attacks on the insured. At one insurance company, a policy change was made and its claims adjusters were instructed not to reveal the address and telephone number of its clients. Some time later, a rear-end accident occurred to an insured's vehicle stopped at a red light. The rear-ender called the claims adjuster and she violated the policy (committed an error) by providing the caller with the insured's home address and telephone number. The rear-ender immediately went to the insured's residence and made threats because his driver's license could be revoked if he had no insurance to cover the claim. A week later the insured came out of her home and found that two tires on one side of her vehicle had been slashed with a knife and destroyed. Apparently, the large insurance company had not conducted an effective audit or field test to determine whether the company policy was properly implemented. Such special testing should be periodic when human error is likely to occur and the consequences could be substantial.

Cautions

Among the general theory-based issues that may be highlighted in error testing are: Are the errors primarily knowledge based or rule based? Are errors significant only because an overly human-centered, user-friendly, and humanitarian bias exists, as opposed to satisfying genuine company needs, or is there a fair balance of interests? Are the errors simply deviations from theoretical scenarios and assumptions of rational operators, without real risk consequences? To what extent are errors forced or shaped by peer pressures and supervisory influence? Are quantitative methods, such as probabilistic risk assessment, more productive and do they provide greater assurance of reduced error and harm than qualitative methods of discovery and correction?

Caveats

Minimum cost

The lowest cost type of testing for human error is to simply piggyback or utilize other scheduled tests, using direct observation and personal analysis. This is relatively unobtrusive, somewhat natural, and seemingly informal.

The difference

The likelihood of human error is most apparent to those experienced, sensitized, and focused on its appearance during test operations. The error specialist has a different perspective than other test personnel conducting tests under stressful and demanding conditions.

Life cycle

Testing for errors should start at concept design and end at ultimate disposal. The aftermarket conditions of use should not be ignored by a failure to test.

Overlap

There is a general overlap between the technical (human–machine) analysis of human error and the management (human–people) analysis of error. There is a community of interest in error reduction for that which is being tested and the error reduction in how the tests are actually being conducted.

Neglect

The testing described in this chapter could be a valuable opportunity to gather information on error detection, cause, and probable remedy. This assumes a timely intervention by an error specialist. In essence, to err is expected; the remedy should not be neglected.

chapter nine

Managerial errors

Introduction

A manager within a corporate enterprise may be the source of human error that negatively affects the company and the employees that he supervises or controls. It is his own personal errors that may come under scrutiny in an error investigation and analysis. The manager is also responsible for the errors committed by his subordinates. Fortunately, this dual responsibility could be well served by the common aspects or features of an appropriate error control program. The manager should, at a minimum, have a basic understanding of human error, its causes, and its control in order to be able to institute appropriate control measures. The key focus of this chapter is error causation, whether manager or subordinate in origin.

Error vector analysis

The following factors should be considered in an error vector analysis to help determine root causation.

Intrinsic factors

There are relatively permanent psychodynamics within an individual that may result in predictable manifestations in the form of conduct clusters (described below) that might spawn error.

Impressed factors

There are bodily conditions that impress upon or overlay an individual's normal or expected behavioral responses, such as prescribed medications and their intended and side effects, diverted prescription drugs and their abuse, and illegal drugs, including stimulants and depressants (as described below).

Extrinsic factors

There are situational stressors that modify behavioral expressions, either increasing or decreasing the likelihood of error. Included are fatigue, heat and cold exposure, hunger, thirst, incentives, diseases, work overload, and discomfort.

Compensatory factors

There may be personal insight, motivation, and some positive or negative capability of change or control of error.

Error analysis

There is a final resultant of the mental or inner (intrinsic) factors, the bodily (impressed) factors, the stressor (extrinsic) factors, and the accommodation ability (compensatory) factors. The resultant is an error vector that forecasts strength (severity of error) and direction (type of error).

A precise analysis may not be required to determine enough about error causation to fashion an effective remedy. Error counting, experimental research, or complicated risk assessment may not be desirable because of cost, time, and the talent required. Well-grounded simplicity can be a virtue.

Error-prone managers

A single mistake (error) by a manager may have far-reaching consequences. Since the manager tells others what to do, mistakes can be easily compounded by subordinates to the extent that the errors may adversely affect the overall viability of the company. The manager is a key person, to whom others look, for decisions or commands that are assumed to be correct and should be obeyed. Not all managers are created equal, have equally critical assignments, or equally manifest the same conduct under the same circumstances.

Some managers may seem to be error-prone in terms of their interpersonal relationships, the role they play in and for the company, and in their own acceptance of responsibility for errors, goofs, mistakes, and the lack of sufficient perfection or the required completeness in performing their assigned job duties. Customarily the manager, when perceived as error-prone, might be moved to a less sensitive job, not promoted or awarded a bonus, or terminated if consistently and substantively below the performance of his peers. This may seem somewhat primitive, but it signals a lack of understanding as to why human errors occur and how they might be remedied.

The term *error-prone* may be disliked, by some specialists, as a disproven concept. Standing alone, such a diagnosis may be insufficient because it generally omits causation. It may be descriptive, but not sufficiently useful. More is needed, as indicated or revealed in this chapter.

When human error problems occur, the reasons should be identified so that there is some basic understanding. An analysis could be made to determine why an error occurred in terms of the predictable conduct of such a manager. Understanding why the error occurred is a first step in deciding between the available choices for an error-prone manager.

A potential manager may be originally selected on the basis of a presumed or predicted ability to play a specific role, in an error-free manner, assuming that error recapture, recovery, correction, mitigation, and countermeasures are available. Similarly, an existing manager may be selected for special training or job reassignment, based on predictable errors in a particular job.

The question then becomes: Is there something more efficacious than the old gut-feeling approach, a reliance on quick subjective impressions, or an after-the-damage management corrective action? The answer is that there are appropriate analytic techniques and organizational remedies.

Analytic techniques

This chapter deals with the conduct that may be displayed by a manager, whether it be a first-line supervisor or a chief executive officer. As used in this chapter, the word *conduct* refers to persistent patterns of human behavior as perceived in relation to established business and social norms.

An illustrative analytic technique, for the assessment of one manager, is the IMED method, as follows:

1. *Identification* of important aspects of needed or expressed conduct. The objective is the isolation of key factors (variables) that could result in unwanted human error.
2. *Matching* the key factors with relevant conduct clusters. The objective is to make a comparison that enables better insight, differential diagnosis, and understanding.
3. *Evaluation* of the resultant categories of expected conduct in terms of functional utility, probability, risk, and general applicability. The objective is to assess the predictability of manifested errors given certain foreseeable circumstances, expected human performance, job demands, social conformance requirements, and the macroenvironmental surrounds.
4. *Decision making* for appropriate action. The objective includes the possible effect of error control, the mitigation of consequences, or an urgent need for virtually complete error avoidance.

Psychological risk assessment

There is an important distinction between the analyses of some managerial errors, described in this chapter, and the technical analysis of human error, as described in earlier chapters of this book. That technical analysis has a focus on the man–machine interface, whereas the managerial analysis has a greater focus on interpersonal problems. While there is a slightly different perspective, they both are systems oriented and deal with human error, its causes, and control. The managerial errors often require a much more detailed psychological risk analysis and a deeper understanding of human conduct.

The conduct clusters, described below, are presented to alert and inform the human error specialist about recognizable atypical patterns of conduct. An understanding and appreciation of the historical experiences of others with such conduct disorders may assist in formulating a realistic and appropriate human error prevention plan that is tailored to the circumstances.

Close attention to error-producing conduct disorders is usually reserved to key afflicted persons or given upon specific request to the error specialist. Otherwise, the time, cost, and diversion of resources may make it prohibitive for the error specialist to get too deeply involved with persons manifesting what may be recidivous conduct.

A psychological risk assessment may be obtained, based on tests, interviews, and a person's history. The assessment may be short or long, informal or formal, with or without follow-up and reference checking, and may be reported orally, in writing, or by computer-generated reports. It should deal with traits and attitudes that are deemed important to a particular situation or circumstance. The risks evaluated might include disruptive alcohol use, rational coping behavior, adaptive capability, anger control, value orientation, rebellious behavior, role adjustment, trustworthiness, conscientiousness, tolerance, social maladjustments, and psychopathic features in an individual's behavior.

Caution: Direct intervention into detailed diagnosis and treatment of conduct disorders should be left to properly licensed psychologists and psychiatrists who are experienced in such professional matters. The error specialist should deal with error situations, not an individual's therapeutic needs.

Conduct clusters

Each of the following *conduct clusters* includes a group of descriptive symptoms, behaviors, or conditions that are somehow tied together, correlated, or associated with each other. Not all symptoms in each cluster need to be present, and each may vary from marginally detectable to mild, moderate, or severe.

The conduct described is easily understood, readily observed in the controlled environment of an industrial enterprise, and described in terms of how things really are in such situations. While qualitative and subjective in character, the descriptors tend to be neurocentric, are useful, and do have predictive value.

Focus (inattentiveness)

A common cause of human error is inattentiveness or a lack of a continued focus. This may be part of a cluster of symptoms reflecting the basic behavioral characteristics of some individuals or groups of persons.

Inattentiveness may be related to an unwillingness to conform to the demands of others (an oppositional defiant disorder), an understimulating environment (boredom), a social relationship difficulty (developmental problems), or very limited personal interests.

Attentional dysfunction is a widespread problem, so common that it is often overlooked or palpably rationalized. There may be neurobiological substrates relative to sustained attention, vigilance, and attentional focus.

If there is deep inattention, there may be a drift into more absorbing, self-interesting, and distracting subject matter. Daydreaming may occur. Inattentiveness may transition to indifference, and there may be a disregard for attention-getting signals and warnings. In other words, what may start as positive attention may decline to a point where it becomes negative.

A state of high human excitation may occur that directly affects the level of human alertness and attention. This may be a constant condition, assuming homeostasis by inverse feedback mechanisms (like a thermostat). This high-stress condition might be maintained for a relatively short time, but then the person may intentionally quench the mechanism in order to minimize the deleterious effects of a continued or constant high level of stress. Alternatively, the human body may resort to an automatic feedback de-excitation in order to return to its normal state, which is a less energy demanding condition that results in a reduced level of alertness and attention. Situational awareness responses are also retarded in the process. There may be problems with divided attention tasks. An individual's attention span may be brief and associated with short-term or working memory deficits.

While there are considerable individual differences as to inattentiveness, consequential errors, and recapture (corrective) mechanisms, it is foreseeable that there may be periodic lapses in attention or some self-protective reductions in focused attention tasks. Better human performance may be achieved by less demanding machine safeguards, by personnel shift rotation, and by ensuring that inattention errors are not critical to the system, or that short attention spans are acceptable.

> *Caution 1*: Induced inattention may result from drug abuse, some medications, a mood problem, or an anxiety disorder.
>
> *Caution 2*: A different diagnostic category is attention deficit disorder (ADD) or attention deficit hyperactivity disorder (ADHD). These may involve a display of inappropriate inattention, impulsivity, and hyperactivity. The inattention may be manifested by a failure to finish tasks, a general distractibility, difficulty concentrating, and just not seeming to listen. There may be associated disorganization, sluggishness, and selective attention problems. This may be associated with

frontal lobe problems, specifically the orbital-frontal areas and the associated connections to the limbic structures. Neurofeedback devices have been used, sometimes together with medication, behavioral therapy, and social skills therapy. There may be side effects from the ADHD drugs that adversely affect the patient's attention and focus.

Inattentiveness, in general, is a complex subject in terms of the many neural systems that could be involved. Generally, it is not a transient event.

Sex (addictive conduct)

Addictive conduct is a common source of human error in industry, commerce, transportation, recreation, and the home. It can occur with alcohol, cocaine, and many other substances and activities that produce compulsive and dependent disorders.

Drug addiction refers to repeated self-administration of a substance that results in a need for increasing amounts of it to achieve a desired level of intoxication. This means there is an ever-higher tolerance for the drug. There are some maladaptive behavioral reactions and unpleasant symptoms (the withdrawal effects). There is a pattern of continued compulsive use (a drug dependence), despite the consequences. Chronic alcoholics are often in denial, believe they are in control, and may have had many unsuccessful attempts to discontinue or decrease the abusive use of the feel-good intoxicants. There may have been full or partial remissions at some time, but there is always a high risk for relapse, and "going off the wagon" often occurs.

Addiction is commonplace. The National Institute on Alcohol Abuse has estimated that up to 18 million Americans have an alcohol-related disorder (Dooren, 2005). Alcohol dependence was defined as four or more drinks a day for women and five or more for men. There are commonplace terms such as *binge* (remaining intoxicated for at least 2 days), *blackouts* (loss of memory for a period of intoxication), the *shakes* (a morning withdrawal symptom), and *hangovers* (a malaise relieved by further drinking).

Addicts manifest a loss of impulse control, an inflexible behavior, a drug-dominated conduct pattern, and an imperviousness to punishment. The loss of control may result from a dysfunction of the prefrontal cortical-striatal system, in particular the orbitofrontal and dorsolateral prefrontal cortex. Alcohol abuse also has been classified as a stress-related psychiatric disorder, affected by life events and maladaptive behavior, and mediated by the central amygdala (Nie et al., 2004).

The cognitive deficits from alcoholism, measured after cessation, are *acute* (impaired intellectual functioning, memory loss, and visual motor skill deficits that can substantially recover over weeks or months), *short term* (impaired problem solving, abstract reasoning, nonverbal memory, and perceptual motor ability, with partial recovery, depending upon the lifetime duration of alcohol abuse), and *long term* (deficits in problem solving, visual learning, memory, and perceptual-motor activities, with slower partial recovery). Heavy drinkers may develop alcohol-induced delirium, dementia, and hallucinosis, exhibit delirium tremens, present amnesic psychosis, have liver failure (cirrhosis), and develop hepatic encephalopathy, which is an impaired liver function that fails to remove toxins that then build up in the blood and cause brain function deterioration.

The undesired behavior accompanying addiction generally involves a reduction in occupational and social activities, including repeated failure to meet role obligations and required work attendance. This often includes excessive time spent on a personal preoccupation on activities necessary to obtain the drug. There may be poor work performance due to hangovers and drug effects. There may be a history of intoxication in high-risk situations, such as machine operation, automobile driving, and recreational activities. There may be marital problems and divorce, verbal aggression and physical fights, and arguments about the consequence of continued substance use and abuse. Some brag, seek the company of other users, and become socially isolated. There may be mood disorders and eventually drug-induced physical and mental disorders.

Cocaine addiction is the pathological compulsive use of the drug. It is manifested by difficulty stopping or limiting the drug use, a continued use despite known harmful consequences, and a high motivation to take the drug (Vanderschuren and Eveitt, 2004; DSM-IV-TR, 2000; Deroche-Gamonet et al., 2005; Robinson, 2004). There is a high likelihood of relapse after withdrawal. Job performance is affected by a strong focus on procurement and consumption of the drug, to the neglect of job duties.

The use of marijuana (hashish, cannabis) is fairly common, and it is a subject of considerable controversy, with strong opinions, among the general public. There are those who believe its use is benign, with helpful medical effects, and those who believe its use is illegal and could result in drug dependence. Its psychoactive effects include brain depression, intoxication (a dreamy high feeling), mood state fluctuations (euphoria with inappropriate laughter), thought disorders (grandiosity and disconnected ideas), a sense of well-being, and lethargy. There may be impaired short-term memory, difficulty with complex mental processes, impaired judgment, communication problems, decreased motor abilities, and possible social withdrawal.

Marijuana abuse produces diverse brain effects, including confusion, delirium, delusions, disorientation, hallucinations, and anxiety reactions. Dependence may be accompanied with withdrawal symptoms, including irritability, nausea, tremors, muscle jerkiness, and insomnia. Continued cannabis abuse is associated with a general apathy, bloodshot eyes, a lack of motivation, poor social relationships, decreasing employment status, and a high incidence of human error. There are tests that can determine marijuana use, even a year later, because it is eliminated from the body rather slowly.

Certain forms of sexual behavior in the workplace may be considered addictive, compulsive, disruptive, and error producing. This does not include flirtations and office romances unless it is a continued behavior that is disruptive of work performance. It does not include the boss's girlfriend, unaccompanied by other problems. It may include an individual's history of continual multiple sexual relationships that distract, create social conflicts, and involve personal preoccupation, reckless pursuits, spreading denial, personal counterattacks, and periods of seeming conformance to norms. The essential feature is uncontrollable addiction that creates opportunities for error of a type that becomes meaningful to a corporate enterprise or other organizations of people.

Caution 1: Not all people who are exposed to addictive agents become dependent or abuse the drugs. Some people just have a higher vulnerability to drugs and may have differences in their brain reward systems. Addiction refers to continual drug-seeking and drug-taking behavior, with such a high desire that it overwhelms normal behavior and any fear of adverse consequences. It differs from social drinking, prescribed medication side effects, and toxic exposure symptoms.

Caution 2: Distinguish between *depressants* (downers such alcohol and narcotics), *stimulants* (uppers such as amphetamines and cocaine), and *hallucinogens* (such as marijuana and LSD). Further distinguish between *adhesives* (airplane glue containing toluene and rubber cement containing hexane, methyl ethyl ketone, and methyl butyl ketone), *aerosols* (spray paint containing butane and hair spray with propane), *solvents* (nail polish removers with acetone, lighter fluid with butane, and gasoline), *cleaning agents* (dry cleaning fluid with tetrachloroethylene), *room deodorizers* (alkyl nitrate and butyl nitrate), and *desert sprays* (whipped cream with nitrous oxide).

Caution 3: Remember that this inquiry is to determine human error causes and methods of error control. Thus, it is important to separate addictions from other substance-related disorders, such as those from heavy metals (lead), poisons (strychnine), pesticides (nicotine), antifreezes (ethylene glycol), inhalants (paint), other toxic substances, and high doses of medication. There may be different effects from different substances in different people.

Caution 4: Do not be surprised by a high frequency of conduct disorders. A report by the National Institutes of Health (Abboud, 2005) indicated that 25% of Americans had a psychiatric disorder. About 60% did not seek treatment. The severity was 40% mild, 37% moderate, and 22% serious. It included 18% anxiety disorders, about 10% mood disorders, and 9% impulse control disorders. This constitutes the general population, from which managers are recruited for various occupations.

Truthfulness (deceit)

The character trait of truthfulness is very important in business practices and in other social activities. Truthfulness helps to establish the credibility of a person and provides justifiability for reliance on that person's acts, signals, messages, and communications. It gradually builds trust, cooperation, reciprocity, and acts of mutual benefit. False or misleading information can quickly destroy credibility and reputation. While truthfulness is desirable, it is a fairly complex matter to assess prospectively and even during its occurrence. As will be explained, too much of the truth, in a rigid form, may not be desirable. A little salesmanship and exaggeration may be needed. But there are many conduct clusters in which deceit, misrepresentation, and falsehoods can be expected at inappropriate times. The human error specialists cannot justifiably rely and act upon information from sources that are not credible and trustworthy in terms of the whole truth.

White lies are considered near-truths, a shading of meanings, close inferences, or slight obscurations. They are intended to maintain civility, forestall immediate embarrassment, stall the need for demonstrated action, or present adverse information in a socially and personally acceptable manner. White lies demonstrate that there are acceptable and, perhaps, desirable departures from complete truthfulness. Several questions arise about white lies. How do you clearly distinguish between white lies and black lies? Is the propensity to tell white lies because of perceived social necessity, a precursor of false information at important times, or because of a perceived business necessity?

There are "weasel words" that are ambiguous and vague, having different meanings to different people. Their use has been elevated to almost an art form by some politicians and marketers. The intent is to seem truthful while avoiding any commitment to specifics. They may be on all sides at the same time. Such words may be misleading to everyone when the result is indecision at a time of needed decisions.

In everyday business practice, there may be great variation in truthfulness. A minor amount may be tolerable, a moderate amount may be distressing and disturbing, and a large departure from the truth may result in angry customers, regulatory fines, and lawsuits. In terms of minor departures from the truth, there may be customs and practices that the whole truth should not be expected, as shown by the following terms and phrases. *Puffing* is usually considered an expression of personal opinion, not a representation of fact, made by someone attempting to sell property. A puffer makes spurious bids in order to run up the price at an auction. In other words, there is an accepted business practice of exaggerating the truth, so that "buyer beware" has become familiar to most people. A promoter or advocate is expected to stretch the truth to convince and sell others on an opinion. A so-called sophisticated argument is that which is clever, skillful, fallacious, not supported by sound reasoning, and devised for the purpose of misleading.

A lack of truthfulness may result from people playing the *blame game*. There may be elaborate attempts at concealment, evasion, subterfuge, cover, and masking to prevent future allegations of unit or personal fault. The potential blamers deflect, revise, reconstruct, or forget in order to avoid possible accusations of fault. This behavior is an early preventative action to obscure the record and prevent future fingerpointing. Documents may be devised that shift responsibility to suppliers and contractors, so there is a neat set of papers where accountability for any future trouble has been assigned to and accepted by another company. Obviously, the confusion engendered by the blame game is not helpful to the error correction specialist. This type of conduct may be considered aversive bureaucracy and unacceptable cultural conduct.

If obvious untruthfulness is tolerated in a company, it may have a compounding effect where the trivial gradually escalates into legally false statements under government disclosure requirements or other regulations. Company downturns may prompt smoothing or augmentation of sales revenues, creation of improper patterns of self-dealing, and the abuse of power to cover shortcomings. There may be convincing excuses or rationalizations of improper conduct, but the seeds may be sown when early instances of untruthfulness are tolerated or even encouraged. Thus, present minor untruthful acts may be a harbinger of the future growth of pervasive major errors.

Malingering is an intentional attempt to grossly exaggerate or falsify physical and psychological symptoms. The motivation may be to avoid work, to feign illness for financial compensation (a gain or reward), to evade findings of fault, to attempt a removal from accountability for actions taken, to serve as an excuse to avoid personal responsibilities, to escape from or avoid unusually disagreeable or high-stress situations, to fake symptoms to attract attention, or to reduce exposure to a dangerous job assignment. It may be part of an underlying antisocial personality disorder or dissociative fugue (an inability to recall important information) as a disguised escape wish fulfillment. A *facetious (sham) disorder* exists where there is no known external incentives or motivation for the exaggerations and falsehoods. The term *negative response bias* may exist where there is some external motivation to negatively bias performance. This includes conditions of nonoptimal performance as compared to what is reasonably expected based on skills and background of the individual. There may be exaggerated effort but poor performance. Subpar performance is also associated with disinterest, depression, fatigue, anxiety, medication side effects, cognitive impairment, and psychiatric disorders.

Confabulation is the fabrication of stories to fill memory gaps or deficits in the process of attempted reconstruction of patterns and context in reactivated encoded memory. The brain attempts to give meanings and causes to partially recalled memories that may suffer from bias or inaccurate perceptions. Usually the person is unaware of the confabulation problem. In other words, the monitoring or checking process, performed by the orbitofrontal cortex (at the base of the anterior brain), fails in its representation-evaluating and behavior-inhibiting capacities. This frontal lobe dysfunction may also be associated with lesions in various parts of the brain that give rise to *spontaneous confabulation*. More extreme is *flagrant confabulation*, the invention of stories from false perceptions, which may result in socially inappropriate behavior.

Intentional falsification is a concern of managers immersed in a tidal wave of unverified information. The question may be who and what to believe in an age of intentional spin and the presence of a few "bad apples" in most large groups of employees. It is well to remember that in early psychological warfare, government agencies divided new information into three categories: white, to be given a favorable spin or interpretation; gray, to be given a neutral or unmodified interpretation; and black, that which should be given an unfavorable spin or revision, as referenced to a geopolitical plan or current policy assessment. A similar public relations spin may be referenced to a company policy manual or current executive directive or recommendation. In other words, there is almost always some overt and intentional attempt to modify the facts and the impact of information.

In scientific endeavors intentional falsification of data does occur rarely, but the replication requirement serves to self-correct errors over time. In legal matters perjury could be punished. There are individuals who construct an identity, live a life of lies, or devise a method of intentionally defrauding others. In a company bureaucracy, the competition for promotion in rank may be duplicitous behavior in regard to self and the others who might be promoted to a more desirable position.

Company managers rarely have conclusive, absolute, or unimpeachable authority on which to base decisions. The manager should have a healthy skepticism in the pursuit of truth, a constant search for errors, a way of making comparisons of unverified information to the logic emanating from past experience, a means of evaluating the ease of fabricating falsehoods (falsifiability), and considerations of the possible motivations of the proponents of patterns of new assertions or interpretations of facts. There is an analogy between the manager's search for truth to enable justifiable decisions and the human error specialist's quest for truth to enable appropriate error reduction. Trustworthiness is king for both.

Compulsiveness (inflexibility)

The obsessive-compulsive personality disorders are important because of their high incidence (1 to 3% of the general population), their greater frequency among high achievers, and their manifestation of both desirable and undesirable behavioral features that are important to industrial managers. The severity ranges from minor to severe in terms of expressed symptoms, with greater human error from the more severe personality disorders.

A person with an *obsessive-compulsive personality* may be methodical, dependable, serious, orderly, reliable, and something of a perfectionist. There may be a preoccupation with rules, procedures, and schedules, with repetitive thoughts and actions. There may be a tendency to be overly cautious, pay too much attention to trivial detail, and repeatedly check for mistakes (uncertainty). There may be time-consuming and impaired decision making, poor allocation of time, an intolerance of mistakes, considerable difficulty dealing with unpredictable events, a reluctance to delegate, and excessive preoccupation with work. There may be inflexibility, rigidity, stubbornness, self-critical thoughts, and either excessive deference or resistance to authority. In essence, there is a preoccupation with order, control, and perfectionism.

A person may have an *obsessive-compulsive disorder* (a neurosis, impulse control, or compulsive conduct disorder). This is a more severe impairment manifested by recurrent obsessions (persistent thoughts, images, ideas, and impulses) and compulsions (ritualistic and repetitive behaviors that are rigid and stereotyped). The recurrent thoughts are intrusive, unwanted, and discomforting. The theme may be risk of harm and danger that does not match actual risks. There may be repeated doubts about things. The obsessions and compulsions are time-consuming, cause personal distress, are known to be excessive and unreasonable, result in anxiety, and interfere with normal occupational functioning. There may be ego dystonia, a feeling that things are not under control or as expected. There is no loss of contact with reality. There may be problems in reading, computation, concentration, and role functioning.

Caution: Distinguish between obsessive-compulsive personality disorders and narcistic personality disorder, antisocial personality disorder, schizoid personality disorder, and symptoms expressed in chronic substance abuse.

Avoidant personality (self-isolation)

Persons with an avoidant personality disorder appear shy, timid, quiet, and inhibited. They attempt to avoid humiliation and personal rejection by avoiding starting anything new, by having a restricted lifestyle, and by declining job promotions or assignments to new projects. They act with restraint, say little, have a tense demeanor, are socially inept, and have a low self-esteem. About 1% of the general population has an avoidant personality disorder. At work, they avoid personal contacts because of an oversensitive fear of personal inadequacy, ineptness, and resultant disapproval, disappointment, shame, criticism, ridicule, and rejection. They are distressed by their inability to comfortably relate to others but want to have relationships with others. The errors are more likely to be those of omission, which may be difficult to correct. Avoidant personality traits are a problem when they become inflexible, persistent, and cause risk intolerance, distress, social withdrawal, and functional impairment.

Caution: Distinguish avoidant personality disorder from social phobias, dependent personality disorder, and schizoid personality disorder.

Dependent personality (indecisiveness)

Those with a dependent personality disorder manifest an inability to function independently. They are indecisive, have a fear of making mistakes, postpone decision making, cannot complete assignments on time, and passively permit others to assume responsibility. They may be submissive, pessimistic, have self-doubt, lack self-confidence, believe they need the help of others, and have difficulty initiating projects. They may have difficulty expressing disagreement with other people because of a fear of offending them. There is an excessive and unreasonable fear that they are completely dependent on others, and they seek dominance from others. There is a pervasive need for reassurance by others. Error of omission can be expected from the indecisive and delayed decision making.

Caution: Distinguish dependent personality disorders from mood disorders, panic disorders, avoidant personality disorders, and changes due to a medical condition.

Social anxiety disorders (phobias)

Those with social phobias have irrational, unrealistic, and intense fear reactions to some specific cued social situation. The fear may be of public speaking, meeting and talking to strangers, signing a document before witnesses, or speaking to authority figures. There may be a sense of possible embarrassment, humiliation, inappropriate actions, or inability to express thoughts correctly in words. This fear provokes an intense anxiety reaction or panic attack. There may be a feeling of impending doom and an urge to escape. There may be sweating, trembling, nausea, muscle tension, and confusion. In some cases, there can be fearful cognition and anticipatory anxiety. Most people have some anxiety reactions, but perhaps 10% of the general population have a serious anxiety problem. The consequence is underachievement at work, defects in social skills, and errors of omission.

Caution: Social phobias should be distinguished from agoraphobia, which is not limited to social situations or scrutiny by others, separation anxiety disorders, schizoid personality disorders characterized by a lack of interest in relating to others, and nonpathological performance anxiety, such as stage fright, which lacks marked distress and impairment.

Social exclusion (pain)

The expression of aversive feelings of social exclusion, isolation, and being left behind is important in the interpersonal, social, or even family-like setting of an industrial or commercial enterprise. It might be just a short-term social separation pain (social loss) that might benefit from social support or a soothing human relationship. More significant are feelings of being shunned, with hurt feelings, social distress, and emotional pain. This type of long-term emotional distress is mediated by the anterior cingulate cortex, located at the brain's midline, and is part of the cingulate gyrus tissue that functions when cognitive tasks are stressful and require close attention. In addition, the right ventral prefrontal cortex responds negatively (Eisenberger et al., 2003; Panksepp, 2003), and other areas of the brain become involved, including the limbic system. The cognitive control for downregulating pain originates in the prefrontal cortex and is relayed through the insula to the appropriate interacting neural networks.

Attitudes and beliefs can shape the very real psychological pain mediated by the anterior cingulate cortex. They can also modulate activity in the pain-sensitive or pain-related areas of the brain. The expectations of pain can be reduced by activity in the prefrontal cortex or by the administration of analgesics or placebos (Wager et al., 2004). In essence, feelings of loneliness, sadness, being ignored, believing there is an exclusion from social activity, and experiencing social pain are important symptoms. They interfere with the ability to form friendships and attachments and to establish social bonds on the job, reduce productive coordination between departments, and could degrade individual cognitive task performance.

Schizoid personality disorder (loners)

This disorder involves a pattern of emotional coldness, aloofness, introversion, and withdrawal from social interactions. A person with a schizoid personality is preoccupied with his own thoughts and feelings, talks little, daydreams a lot, selects solitary activities, and prefers social isolation. He copes by fantasy, believes the fewer activities, the better, and is indifferent to praise, criticism, or the feelings of others. The schizoid personality may have difficulty expressing anger, have a passive reaction to adverse events, have goal drift, and is detached from social activities. There may be a significant impairment in occupational activities, social skills, and error capture.

Caution: Distinguish Schizoid personality disorder from delusional disorder, autistic disorder, paranoid personality disorder, obsessive-compulsive personality disorder, and changes due to the effects of a medical condition on the brain.

Paranoid personality disorder (suspiciousness)

A person with a paranoid personality has an evasive suspiciousness and distrust of others without justification, doubts the trustworthiness of others, attributes malevolent motivations to other persons, and reads hidden meanings into benign statements. Those with the disorder are stubborn, difficult to get along with, bear grudges, and do not forgive slights or insults. They are rigid, critical of others, unable to work collaboratively, attempt control over others, and voice sarcastic expressions. The prevalence of paranoid personality disorder is about 1 to 2% of the general population. The paranoid personality is reluctant to confide in others, tends to misinterpret communications, is always attentive to possible harm from others, and is quick to react with anger, argument, and counterattacks. They are often underachievers, become involved in legal disputes, and have negative stereotypes of others. If they work in isolation, they may be efficient if the work involves simplistic assumptions. There may be occupational problems from their combative and suspicious attitudes. They are of concern in regard to human error when these traits are inflexible and maladaptive.

Caution: Distinguish paranoid personality disorder from delusional disorder, avoidant personality disorder, antisocial personality disorder, and chronic substance abuse.

Passive-aggressive personality disorder (negativism)

Individuals with this disorder are prone to procrastination, contrariness, ambivalence, defeatism, inefficiency, sullenness, and are passively resistant to demands for adequate job performance. Their concealed hostility and aggressiveness may become manifest in a subtle undermining of work tasks and projects by intentional inefficiency, argumentativeness, forgetfulness, and inappropriate assertive behavior. They resent and oppose demands by authority figures to function at a level expected of others. Examples of errors to be expected are misplaced materials, misfiled documents, and other errors of intentional commission. They may chronically complain of being misunderstood and unappreciated, are envious and resentful of others, and complain of personal misfortunes and their poor future prospects. Their indecisiveness and erratic performance may forecast a variety of human errors.

Caution: Distinguish passive-aggressive personality disorder from oppositional defiant disorder, dysthymic disorder, and major depressive episodes.

Depressive personality disorder (sadness)

This affective or mood disorder is reflected in a feeling of profound and intense sadness that persists beyond that reasonably expected. The disorder may also be known as depressive neurosis, and sometimes as a dysthymic disorder. It is characterized by feelings of gloominess, sadness, broodiness about past events, pessimistic attitudes, feelings of dejection, joylessness, unhappiness, loss of interest in usual or enjoyable activities, social withdrawal, and chronic tiredness. The depressive personality may seem to be overly serious, lacking a sense of humor, always anticipating the worst, and unassertive. The individual may have beliefs of worthlessness, inadequacy, remorsefulness, low self-esteem, and proneness to guilt. There may be decreased concentration, attention, clarity of thought, and work productivity. The errors expected are those of omission.

Caution: Differentiate depressive personality disorder from a substance-induced mood disorder, such as alcohol dependence, an obsessive-compulsive disorder, a major depressive disorder, and a short-duration sadness.

Posttraumatic stress disorder (anxiety)

This anxiety disorder can be originally initiated by exposure to an overwhelming traumatic event that provokes feelings of horror, intense fear, and helplessness. The event may be the threat of death or serious injury, disturbing events of military combat, robbery, mugging, violent sexual assault, life-threatening illness, disasters, and serious accidents. Subsequently, a stressor stimuli associated with the event may arouse or may trigger an intense reexperiencing of the trauma, flashbacks, recollections, or nightmares in a particular or characteristic pattern.

The onset and duration may be acute, chronic, or delayed. This is generally followed by an avoidance of stimuli such as thoughts, activities, events, or people that could arouse or trigger the stress reaction. The prevalence of posttraumatic stress disorder is about 8% of the general population and 30% of soldiers returning from combat. There may be an emotional numbing and deliberate attempts to avoid reminders of the trauma and its consequences. The result may be prolonged psychological distress, impairment of occupational activities, physiological reactivity to cues of the traumatic event, feelings of detachment from other people, restricted emotional feelings, hypervigilance, irritability, difficulty concentrating or completing work projects, and errors of omission.

Caution: Distinguish between posttraumatic stress disorder and an adjustment disorder with an anxious mood, differing response patterns, and, perhaps, a less severe stressor; an acute stress disorder that is short lived; an obsessive-compulsive disorder not related to an extreme traumatic event; and malingering or feigning of symptoms.

Antisocial personality disorder (disregard)

The characteristic personality traits of people with an antisocial personality include a pervasive pattern of callous and cynical disregard for the rights of others, an exploitation of others for material gain and self-gratification, the constant use of deceit and manipulation, a lack of empathy for others, an impulsive acting out, and serious violations of rules without remorse or guilt. The prevalence of antisocial behavior in the general population is about 2%. This conduct has been called psychopathic, sociopathic, and dysocial. The antisocial individual demonstrates a continual disregard and contempt for the feelings of others, there are repeated attempts to con others, and there is general irritability with an unusual aggressiveness in words and conduct. The person may be excessively opinionated, arrogant, and verbally facile to outwit others. Antisocials tend to be thieves, racketeers, drug dealers, vandals, and openly willful violators of almost any rules. They demonstrate irresponsible work behavior, repeated unexplained absences from work, and minimum remorse for the consequences of their behavior. In general, there is a failure to conform to social and legal norms, repeated lying, dishonesty, and blaming others for their transgressions. There is a characteristic reckless disregard for their own personal safety and the safety of others. Their antisocial behavior is rarely modified, and committing intentional human errors is an attractive option for such individuals.

Caution: Distinguish between antisocial personality disorder and narcissistic personality disorder, histrionic personality disorder, paranoid personality disorder, schizophrenia, and substance-related disorder.

Other disorders

The basic purpose in describing some of the various common personality disorders is to illustrate a typically forgotten or ignored source of error causation and control. In addition to those disorders already described, there are other similar root sources of error. *Histrionic personality* (hysterical) is where emotions are exaggerated, contrived, dramatic, enthusiastic, childish, and often hypochondriacal. It is attention-seeking behavior, and if the person is not the center of attention, he or she feels unappreciated. The motivation is a wish to be protected, dependent, and have others solve their problems. In occupational settings, the behavior is excessive, the speech is superficial, strongly expressed opinions may not have supporting facts, emotions are on public display, and they are overly trusting and highly suggestible. *Narcissistic personality disorder* (superiority) is where there is grandiosity, an exaggerated belief in self-importance, self-value, and personal superiority. Persons with this disorder have a need for constant admiration, inflate their accomplishments, and are boastful, self-centered, selfish, arrogant, lacking in empathy, and pretentious. They expect to be given whatever they want, have a sense of entitlement, and are exploitative of others. *Bipolar disorder* (mood fluctuations) is where there is an affective disorder characterized by alternations between manic and depressive episodes.

Note: This information should encourage collaboration with licensed professionals, such as clinical psychologists and psychiatrists, who can provide credible and accurate information on diagnostic classifications and their possible contribution to human error in a particular set of circumstances.

Overlays

Drugs

It is a fact of life that a significant proportion of the general population is taking or has taken drugs that alter normal brain functioning. This includes controlled prescription medications with behavioral side effects, under-the-counter prescription medications that are illicitly diverted to persons for substance abuse, and a wide variety of illegal mind-altering drugs. Their behavioral effects may be as depressants (downers), stimulants (uppers), hallucinogens, sedatives, or some other significant central nervous system (CNS) function alteration. There may be a psychological dependence on drugs to induce pleasurable mood changes such as euphoria, to relieve tension and anxiety, to avoid discomfort, and for sensation alterations.

These drug use effects are superimposed as an overlay on an individual's basic brain structure function, personality attributes, and psychodynamics. In essence, the person is changed and the manifested propensity for error is altered. A few of the more common drugs will be described for illustrative purposes.

Psychostimulants (mild central nervous system stimulants) may be addictive when continually used to lose weight, to produce euphoric highs (short-term mood elevations), or to stay awake for prolonged periods (all night by college students). The psychostimulant drugs include amphetamines (known as speed or uppers), methylphenidate hydrochloride (Ritalin), and pemoline (Eylert). These drugs affect the release and uptake of dopamine and norepinephrine (neurotransmitters), or they may activate the brain stem arousal system. *Ritalin* is used in the treatment of attention deficit disorders (minimal brain dysfunction), narcolepsy, and convulsive seizures. Dexmethylphenidate (Focalin) is another mild CNS stimulant used to treat attention deficit hyperactivity disorder. The use of psychostimulants may be evident in graveyard shifts, during the preparation of engineering progress reports, or in formulating marketing proposals in the course of preparing and selling important marketing proposals. During the subsequent crash or letdown (the abstinence syndrome), the sluggishness and depression may result in a high frequency of errors.

Cautions: Drug use and dependence are matters that may involve fundamental rights to personal privacy, medical privileges and confidentiality, employment discrimination regulations, and the professional practice of medicine. Coordination and collaboration with a licensed physician and a law specialist is advisable for any actions other than obtaining a better understanding of error causation.

Substance abuse disorders may be continuous and maladaptive, episodic for short periods of maladaptive use, or in remission either during treatment or for some long duration of time. These classifications may be quite uncertain. The term *abuse* refers to the dysfunction caused by a drug.

Oxytocin decreases anxiety and stress, facilitates social encounters, and extinguishes avoidance behavior (Huber et al., 2005). It calms the body, reduces depression, increases trust, promotes social bonding, and increases the pain threshold. Oxytocin is produced by the nerve cells within the hypothalmus at the base of the brain and by other cells in the body as neuropeptides. It is released by axon projections into the posterior pituitary gland, the master gland, and circulates in the blood as a neurotransmitter. Synthetic oxytocin is known as pitocin, syntocinon, and uteracon. Oxytocin modulates the autonomic expression of fear by extinction processes. This mind-altering drug may cause errors of commission.

Valium is one of the most common drugs used to treat anxiety disorders. It is a nervous system depressant that promotes mental and physical relaxation by reducing brain activity. Valium's generic name is diazepam, which is a benzodiazepine derivative. There are eight benzodiazepine drugs on the market, and each could result in a physical dependency problem. They act on the limbic system, the thalamus, and hypothalamus to produce a calming effect. It is also used to treat convulsive disorders. The side effects include drowsiness, fatigue, ataxia, and confusion. The errors are those of omission.

Opioid analgesics are pain relievers related to morphine. They are narcotics with many side effects, including addiction. They include codeine, methadone, and seven other drugs. The most commonly abused opiate is heroin, with 800,000 people chronically addicted and 250,000 on methadone used as a detox agent. Methadone, a synthetic opiate, is used to block the brain's opiate receptors in order to reduce the craving for heroin or morphine. OxyContin is a synthetic opiate of growing importance. The symptoms of these narcotics include drowsy mindlessness after a euphoric rush, then total preoccupation with securing another fix or complete passivity if another shot will become available. Such narcotics are bad news since a heroin habit can be costly and the money has to be obtained somewhere. Human error is to be expected, but capture and correction by the individual are moot.

Harmful chemicals (chemophobia)

The industrial manager should be aware that exposure to harmful chemicals could result in increased human error from neurological impairment.

A 2005 biomonitoring report from the Centers for Disease Prevention and Control, based on blood and urine samples from 2400 people who were tested for the presence of 148 chemicals, found that the presence of harmful chemicals is common. Approximately 76% of the population had pyrethroids, a pesticide used as a roach killer, in their blood. Three types of phthalates, used in cosmetics and plastics, were found in 5% of the population at levels above those associated with genital abnormalities in boys. There were traces of chloropyrifus, lead, cadmium, and cotinine. The cocktail effect of trace amounts of several chemicals in combination, whether cumulative or synergistic, is presently unknown.

There have been indications that very low levels of industrial chemicals might be involved in a variety of cancers and brain disorders (Waldman, 2005). Even background levels and no-observable-effect levels may promote gene activation and alter brain development. This includes learning problems in children from mercury, altered brain structure and behavior from bisphenol A, and thyroid or behavior changes from perchlorate. There may be no safe exposure levels for lead and some other substances. Japan has designated 70 chemicals as endocrine disrupters that interfere with hormonal signals. Some hormones, such as estrogen, can trigger biological responses leading to brain abnormalities. The adverse chemical effects may be arguable and could be resolved one way or the other, but they deserve current consideration by the industrial manager concerned about human error.

Denial

Denial of the harmful effects of drugs and chemicals is common among those exposed. Many cannabis users strongly deny that it could cause brain injury. The evidence is otherwise for this mind-altering drug. For example, smoking cannabis releases endocannabinoids by the pyramidal cells and interneurons in the hippocampus, and they affect the normal brain development of the fetus and newborn child (PRS, 2005). In terms of retaining employment, there is strong motivation to deny drug use or the harmful effects of the drug.

Cigarettes

There are many heavy cigarette smokers who are aware of, but deny, its adverse health effects. They may be self-medicating with nicotine to achieve an increased feeling of pleasure and satisfaction resulting from the increased mesolimbicocortical dopaminergic activity. This is particularly true where there is an underlying reduced neuronal activity in the frontal cortex of the brain.

Coffee

The widespread use of coffee also suggests a denial mechanism. Coffee contains caffeine, a psychoactive substance that has a stimulating wake-up effect. Up to 90% of Americans consume some coffee every day. Additional caffeine comes from tea, colas, milk chocolate, Anacin for headaches, Vivarin, No Doz, Dexatrim, and other products. Caffeine increases dopamine levels, producing pleasure, and adrenaline (epinephrine) is secreted by the adrenal glands, producing the fight-or-flight response.

After the caffeine wears off, there may be anxiety symptoms, depression, fatigue, tiredness, a jumpy and irritable feeling, and stress reactions. Caffeine results in a general modulation of central nervous system functions. It is sometimes used for weight loss, and the liver releases sugar into the blood-stream for extra energy. It is considered addictive and capable of abuse. Caffeinism, chronic excessive caffeine consumption, leads to anxiety, depression, and degraded human performance.

Caffeine produces persistent blood pressure increases, and lifelong caffeine use may contribute to cardiovascular disease (James and Gregg, 2004). Caffeine under active stress coping increases heart rate and decreases ventricular ejection time (Hasenfratz and Battig, 1992). Caffeine has a CNS activation increase under low mental loads, but stimulates errors under medium and high mental loads (Navakatikian and Grigorus, 1989).

Six million people work a night shift, about 10% of whom are excessively sleepy at work and have difficulty sleeping during the day. This shift-work sleep disorder increases human errors associated with mental alertness, reaction time, and attention. There are increased near misses and drive time accidents. The use of caffeine is a common antidote to sleepiness at work. There is also business travel resulting in jet lag (circadian dysrhythmia) and attempts to use caffeine, alcohol, and melatonin (the hormone that regulates the sleep–wake cycle). Caffeine is commonly combined with other drugs.

The denial of harmful health effects from drinking beverages containing caffeine is widespread and may suggest why caffeine ingestion is both socially acceptable and unusually popular. Few coffee consumers believe that it could be addictive despite what they may see in others.

Note: There may be off-label (unapproved) uses of drugs, investigational uses of drugs, the use of atypical drugs, or clinical trials involving placebos. Depending on the understanding of the patients, there may be a denial of any specific drug or medication use. Any comorbidity would be confusing except to the drug provider.

Caveats

Confirmation

The preliminary evaluations made in accordance with the contents of this chapter should be considered first impressions. If a neutral or positive assessment is made, this may be within the realm of the experience of a nonspecialized but experienced professional. If it is a negative assessment, no significant decisions should be made without supplemental information and specialist opinions that could confirm, condition, deny, or extend the first impression.

Supplemental information

A wide range of information is currently available on many relevant topics. For example, the side effects of medications, including behavioral manifestations, are described in the ethical or prescription drug's information sheet, its package insert, or the annual *Physicians' Desk Reference* (book or electronic version), published by Medical Economics Data Production Co., Montvale, NJ. Considerable supplemental information on many topics, including biographical data for key people, is available on the Internet.

Confidentiality

Oral history interviews, employee surveillance material, organizational audit or screening data, and other personal evaluations (derived from the methods described in this book) could contain information that might be considered an invasion of privacy. Even with an informed consent document, such material should not be disseminated (republished) or made available in written form where others might see it. The details of the evaluation should remain confidential and not become a matter of ethical oversight (for further information, see Chapter 12).

Intervening causation

There may be other physical and neuropsychiatric causes for the conduct and behavior anomalies that are undesirable or unacceptable in the workplace. For example, the early-symptom, neuropsychiatric, adult-onset neurodegenerative diseases include Alzheimer's, Parkinson's, and Huntington's diseases. Also included are fronto-temporal dementia, motor neuron diseases, schizophrenia, and bipolar disorder. These are associated with various genes and their mutations. They cause progressive neuronal degeneration. There may be biomarkers for early disease symptomatic diagnosis and possible disease-modifying medications or immunotherapy options available to the medical treatment specialist. That specialist should be consulted as to the degree of intervening causation in job performance as well as prognosis.

Rapid changes

Current neuroscience is rapidly progressing because of advances in neuroimaging techniques, neurocognition, neurochemistry, genetics, and molecular biology. Discoveries mean more information, more specifics, more understanding, and more changes. Thus, the information presented in this chapter may have present value, but will soon be outdated and its value diminished to some degree.

Do no harm

The kind of information generated in these studies may be harmless to some and toxic to others. Avoid any potential causes of pervasive harm that conceivably might accrue to others. A high degree of special care is needed. There should be a conscious effort to do no harm to others.

Secondary gain

The techniques that could be utilized to reduce and control managers' errors are goal-directed inquiries of a very personal sort. They should be conducted with sufficient care, sensitivity, finesse, and skill that a friendly rapport develops. The manager could believe that the company is taking a special interest in his words, giving him some recognition in a personal manner, and determining how to provide some guidance on how to do even better personally and at work. This kind of favorable interpretation and personal reaction may encourage company loyalty, dedication to work, and, as the Hawthorne and related studies indicate, greater productivity and job satisfaction. In essence, handled properly, the techniques should produce both primary goal benefits and secondary gain.

chapter ten

Institutional and organizational errors

Introduction

There are human error problems that go far beyond the simple errors attributed to one or two managers (*managerial errors*). Pervasive human error problems may afflict one company (*organizational error*) and result in loss of market share, financial viability, customer satisfaction, and adaptability to competitive forces. The human errors may spread over a group of companies or a trade group and can be considered *institutional errors*.

Institutional errors

Consider the "sludge problem" that occurred in some automobile engines during the years 1997 to 2005. Oil sludge is a thickened oil that fails to properly lubricate the engine parts. This may cause the engine to stall or stop unexpectedly, damage the engine, and necessitate an expensive engine replacement. At first, it was considered a vehicle *owner error* (poor maintenance) because the oil was not changed frequently enough (neglect). The owner's manual recommended an oil change schedule, but driving a vehicle in dusty or extreme conditions may require triple the number of oil changes. The engines were also failing during normal service, but warranty claims were often rejected. *Dealer error* was alleged because claims were mishandled or arbitrarily rejected. The fact remains that only certain engines in certain vehicle brands suffered the premature failure.

This could not be the result of managerial error (one mistake) or organizational error (company misjudgment) because the sludge problem occurred in vehicles manufactured by four different companies and affected about 5 million engines of six different types. The problem was persistent, lasting 4 to 8 years of production (1997 to 2004). Various reasons as to the cause of sludge have been given (Rechtin, 2005), including hot spots that bake the oil, cold spots that create acid and sludge, and poor engine breathing that permits combustion gases and oil vapor to form sludge.

There have been many allegations of blame involving consumer error, dealer error, regional service representative error, manufacturing error (tight fit among the moving parts of the engine), and design error (accommodation of technical advances and relocation of parts). Despite the finger-pointing, general questions remain. In the competitive rush to market the engines and vehicles, was there sufficient premarket durability testing, appropriate consideration of consumer oil change habits, and a customer satisfaction approach to warranty claims and informal recall repairs? If there is common fault among various companies it is institutional. This chapter explores some institutional and organizational human errors, their cause, and the applicable countermeasures.

Common viewpoints

The December 7, 1941, attack on Pearl Harbor has been called a disaster because of the alleged personal errors of the two commanders-in-chief in Hawaii. It has been claimed that Admiral H.E. Kimmel failed to have his long-range reconnaissance aircraft patrol in the location used by incoming hostile aircraft. His counterpart, General Walter E. Short, assembled his aircraft together in a group to avoid a threat from saboteurs (Allard, 1996). Was this a double human error for which personal blame was warranted? Could it have been an organizational or military system error? There were attitude and perspective problems created in the military services by the belief that surface forces were not particularly vulnerable to aerial attack in the middle of the Pacific Ocean, since airpower was more theory than reality. There was no sense of protective urgency by the on-scene commanders despite ominous diplomatic and intelligence assessments in Washington, D.C. There were problems in unity of command in the form of service sovereignties and organizational autonomies. It was the system that failed to provide appropriate information (intelligence) to alert the on-scene commanders to a present danger and to overcome the limitations imposed by the structure and function of the overall organization. Common viewpoints that are in error suggest an organizationally induced error.

Mind-set

Engineers may be limited in their willingness to accept new ideas for new projects, regardless of the source of those ideas, be it from trusted suppliers or from the maturation of new materials and processes. In defense, the engineer may personally appreciate the success of past designs that evolve ever so slowly through the gradual elimination of troubles and difficulties. The project engineer may fear further cost cutting as possibly compromising the integrity of a well-tested and accepted product. However, this conservative mind-set may be interpreted as a lack of an open mind, excessively narrow thinking, the avoidance of proactive continuous product improvement, an inability to grasp business decisions, a refusal to help manage cost reductions, and someone set in his or her ways.

A too rigid mind-set may be the result of perceived company policy, knowledge of technological evaluations that are accepted by industry counterparts, and a form of compliance with acceptable peer beliefs. In essence, the mind-set may have been the result of a type of indoctrination by others in the organization. If something else is desired, the adverse mind-set must be unwound and a favorable type of open mind should be reset in accordance with company objectives.

The delay by some companies in developing gasoline–electric hybrid vehicles may seem to be an organizational judgment error, a strategic blunder, or a public relations misstep (Truett, 2005). But a general mind-set was created by repeatedly publicizing a more advanced technology, a hydrogen-based propulsion system. There was a detrimental mind-set created, and it served to make engineers ignore a so-called interim technology, the hybrid engine. As might have been expected, competitors saw a weakness in a mind-set-created laxity and pressed their advantage. Since this was a high-level error, on a topic known and reconsidered over a prolonged period and of vital interest to each of three major companies, it should be considered a costly mind-set institutional error.

Perceptual errors

A pharmaceutical company expanded its product mix by forming partnerships, alliances, joint ventures, and acquisitions. One such joint enterprise became a dysfunctional relationship because of suspicions and distrust. Issues discussed by telephone quickly escalated into emotional conflicts. The *errors of personal perception* were pervasive and shared by both organizations. It was decided to get the key protagonists from each company to meet face-to-face at a neutral location. After many complaints were aired, a discussion of shared goals was conducted, and personality tests were administered by an outside consultant. It was concluded that one company was acting in an impulsive manner and the other was acting with a lack of focus, some arrogance, and a fear of mistakes. There were explanations why each company believed as it did and what it was actually doing. There was a serious attempt at conflict resolution, creation of amicable interpersonal relationships, and scheduling of separate and joint activities. In essence, joint relationships between high-technology firms, employing professional level specialists, are akin to marriages between strong-willed people. Conflicts are to be expected, resolution of differences can occur, and fruitful outcomes are a reasonable goal. Personal perception errors can be modified to agree with reality and joint venture needs.

Reciprocity

A company may function as single integrated organization, as a headquarters with different divisions or controlled subsidiaries, as a collection of independent organizations, each responsible for its own actions, or in some other format appropriate to its business plan. The identification and control of organizational human error becomes more difficult with functionally disparate entities, located in different countries, and without uniform standards of policy, practice, or procedures. Diverse corporate entities may share a common brand name and little else. There may be no common ownership or control, no common company culture or country social culture, no common liability requirements, and no common worker or trade union rules. Decentralized, but identical, human error programs are usually avoided as indicia of management control over the entities receiving such guidance. This does not mean that there cannot be shared experiences on an oral basis among visitors to a firm that has a good program. This kind of reciprocity most frequently occurs in committee meetings of standards associations, trade groups, or professional societies.

Organizational human error may occur in just one independent company. It should not be dismissed as something unique to that one company. A means should be devised, that is legally acceptable, to dig deeply to extract the lessons learned. Organizational human error can be so costly and disruptive that any meaningful learning experience should be rapidly exploited for the benefit of others. Reciprocity suggests a benefit to all participants in the professional endeavor of the discovery of true cause and effect.

Open loops

Unqualified employees may commit a disproportionate amount of human errors in their trial-and-error learning efforts. The Peter Principle suggests that those people, in business organizations, rise to their level of incompetence. In flat organizations (just a few layers of management), this source of organization error may become quickly apparent. In larger and more complex hierarchical organizations, the error manifestations may be latent, subject to fault transference, or attributed to subordinates or other employees. The incompetent individuals may become difficult people to deal with because of their self-protective behavior, transference attempts to so-called unreliable subordinates, and their conduct, which seems contrary or cantankerous.

These incompetent employees are open loops in an organization's communication and performance circuits. What gets by them or performed by them can be so distorted as to lead to organizational error. They may convince other companies, in a trade association, to take actions or inactions that can initiate institutional error. The question may be how to get such dead-wood individuals out of the loop or located in a position where no serious or uncorrectable harm can be created. Generally, such individuals are known, may by well entrenched, and are not taken seriously until another organizational error is created by them. They deserve special attention by human error specialists dealing with organizational and institutional error.

Motivation

Another form of organizational error is the *motivational human error,* where excessively large monetary rewards are intended to promote some desired performance result. If the incentive is very high, it may serve to motivate unethical, inherently destructive, and even unlawful behavior. The high reward fosters self-interest over company or social interest. Reasonable incentives motivate reasonable behavior that is competitively and cooperatively acceptable, rather than behavior that is extreme and unacceptable in hindsight. Unreasonably low incentives seem to result in a general nonthink mentality, lemming-like behavior, and a passivity that is undesirable in the long term. Since establishing appropriate motivation is an organizational responsibility, errors flowing from improper motivation are organizational human errors.

There are those who will engage in cutthroat behavior if minimally motivated by self-interest. Other people will knowingly inflict injury on others, if an ostensible authority directs them to do so. Both behavior modes are unacceptable and should be controlled by stressing social and ethical norms prior to any manifestation of this form of *injurious human error.*

Theoretical risk

A safety test was scheduled at reactor 4 of the Chernobyl Nuclear Power Station. The power output was to be lowered to 700 MW, according to regulations, by lowering the boron control rods into the water-covered uranium core of the reactor. The deputy chief engineer who was in charge, in control room 4, decided to lower the output to 200 MW despite the objections of crew members reminding him that they had been instructed in their training not to go below a 700-MW power limit. When the test proceeded, despite some objections, to the lower level of power output, there was not enough steam pressure generated to operate the electrical power generators, so a switch was made to backup diesel generators. A short delay in diesel-generated power adversely affected the water pumps that supplied coolant water to the reactor. When a power-up was initiated, a water shortage alarm was ignored because everyone believed that the reactor design was safe and there was only a very low probability of an accident. They did not know that a hot spot had built up at the bottom of the reactor and was not detected by the sensors in that location. When the power-up did not go as planned, they decided to drop the control rods, anticipating a drop in power output, but the first contact of the rods with the water created a power surge. Without enough water, steam pressure increased quickly, there was a blast, and then a major explosion. The worst nuclear explosion in history occurred. A 20-mile zone of exclusion was instituted that included the evacuation of the nearby city of Pripat. A great many people were brought in to the plant location in an attempt to control the situation. Many of those died or suffered severe radiation injuries. A radioactive dust cloud moved over much of Europe. Subsequently, there were thousands of cancer deaths.

The cause might be attributed to managerial human error, that is the intentional violation of regulations by an arrogant and rude engineer in charge. There were many accusations of design defects that apparently had been covered up in a rush to construct the plant, including unsafe conditions at low-power-output operations. However, there was an overriding organizational human error, the pervasive belief that the low probability of an accident was only a remote theoretical risk, not an actual risk, because they were convinced of the high degree of safety built into the power station. The engineer in charge did not believe he was taking an unreasonable risk, the crew members in the control room eventually went ahead with the test, and those who scheduled the test did not believe it posed any unusual danger. In fact, the chief engineer was asleep at his home during the scheduled test.

Leadership errors

Most organizations are patterned after the Prussian (Clausewitz) military model, which involves a command-and-control structure. For unity of command, there is generally only one executive office with a designated leader (a president, director, or general) who operates with assistants or advisors (a staff, cabinet, or council). Orders are issued to subordinates (vice presidents, function directors, chiefs, or commanders) who are responsible for carrying out the orders and achieving certain performance objectives. They may also deal with administrative and personnel matters. An important function is the necessary coordination of their authority and delegated responsibility within their jurisdiction and between other organizations. There are successive levels of a branching hierarchy with similar subordinate functions. This autocracy with extended control continues until it reaches the worker level (immediate supervisors, lead engineers, technicians, associates, or soldiers).

There are many variations in the command-and-control structure or organizing concept. Some are overly strict, as in military or police operations, and they may demand strict obedience, no questions asked, and quick stereotypical responses. The correction of specific organizational errors in strict police operations has been proven to be a rather costly process, sometimes lasting 1 or 2 years. Strict by-the-book control and automatic or habitual responses are difficult to modify or replace. Other organizations have comparatively loose command-and-control structures and functions, as in academic institutions or software companies, where individuality and creativity are prized. Error correction may be quicker in loosely organized enterprises, but may also be difficult because nonuniformity is acceptable.

There is an inherent conflict between the unity-of-control efforts of a headquarters or command center and the efficiency and effectiveness of functional decentralization. The command center may be conceived as a coordination, standardization, motivational, synergistic, technical backup, and surveillance or intelligence fusion operation. There may be a bureaucratic bargain with decentralized or quasi-independent operations that believe in essentially self-governing operations that enjoy quick local decisions, virtuosity, flexibility, and close customer support. They argue the concept of decentralized execution, appropriate asset control, and a better perspective of the competitive battlefield. In such situations, some adversary jousting for dominance may occur and perceptions of organizational error become the ammunition in a territorial battle. Fortunately, the errors may be illusionary, self-correcting after a tactical encounter, or of minor import outside the political battlefield.

The general benefits of a command-and-control structure include extended control of a variety of activities or functions, the ease of quick communications, a known collective direction aided by documented business plans, an inherent adaptability and resilience to change, an ability to sustain losses through ease of employee replacements, and simplification of jobs by a narrowing of the tasks to be performed. However, the complexity of operations and an excessively narrowed specialization of jobs create more opportunities for organizational errors, both obvious and covert. Error detection and correction become largely customized to the unique character of each organization.

Leadership errors, from intentional self-interest or poor judgment under the circumstances, are usually correctable when discovered. They may be detected by an oversight committee, an external audit entity, a stockholder group with legal rights, some independent board directors, the public media's investigative reporters, or those exercising government regulatory control. These checks and balances may or may not be effective in the control of leadership errors in complex organizations. Error detectives within a company may be needed.

There should be carefully defined, documented, and established policies and procedures for the organization that should result in predictable performance and results. They are one part of a measurement system for judging organizational performance. If the results do not meet expectations, there may be a need for unpleasant explanations that may generate calls for correction or change. The failure to meet such performance standards may be caused by specific leadership errors that pervert or debase the fundamental purposes of the command-and-control structure.

A key ingredient in the function of the command-and-control system and its participants is mutual trust based on the benevolent reciprocity of past conduct. Perceived malevolent, unjustified, unreasonable, or unfair actions have an unfavorable effect on trustworthiness beliefs over time. When command orders are communicated, the recipient's brain reward pathways react quickly to process the information. The brain predicts logical outcomes on the basis of past trustworthiness, directs appropriate responsive human social interactions, and the result may be fully committed compliance to the order, cautious compliance and partial reactions, or just *noncompliance errors.*

Organizational culture

Assumptions

There has been a prevailing assumption that a good organizational culture will serve to limit human error, and it could actually assist in error reduction efforts. In addition, it is assumed that a good culture provides the kind of group harmony that welcomes a climate of change. As a result, there have been many attempts to better understand worker attitudes, traits, beliefs, and motivations. The culture being assessed applies to all workers in a company, and therefore is considered organizational in character. It is the company managers that create and maintain a particular organizational culture.

Historical perspective

Early attempts to improve manufacturing productivity utilized time and motion studies applied to the elements of simple jobs. The industrial worker was assumed to be only a wage earner who could tolerate poor working conditions as long as he was paid to do so. In the 1930s, the Hawthorne studies (Roethlesberger and Dickson, 1939) suggested that workers formed human social groups with informal leaders to advance the interests of the group. In addition, the Hawthorne effect indicated that workers that were singled out, given attention, and shown consideration would increase their production (a placebo effect).

Work incentives expanded from only pay increases to other motivators, such as good working conditions, social group processes, and management interest in the workers' activities. Later, a hierarchy of human needs (psychological deficiencies) was proposed (Maslow, 1943) that started with the basic needs for survival and moved upward to safety needs for security, belonging needs or love, esteem needs or status, and, finally, self-actualization or self-realization.

Traditional management practices were described (McGregor, 1960) as assuming that workers disliked work, that people must be threatened with punishment to work toward organizational goals, and that people prefer to be directed and controlled. However, McGregor contended that work could be a source of satisfaction, the worker could exercise self-direction and seek responsibility, and creativity is widely distributed in the general population.

The emphasis changed to a culture of meaningful work by job enrichment and the opportunity for workers to manage their own work. It recognized that there is a need for individual achievement and recognition. It reasoned that personal and organizational goals and values should be compatible. It suggests that the use of authority should be minimal. The industrial manager's role is to maintain and improve that form of organizational culture in which good things spontaneously happen.

Industrial engineering

Industrial engineering has been mostly concerned with manufacturing methods improvement, work measurement, administration of wage payment systems, cost control, quality control, and production control (Maynard, 1971). Other definitions include the integration of men, materials, and equipment in manufacturing applications. In fact, industrial engineering has a broad scope and is expanding as modern technology creates new needs.

The corporate manager should be well aware that management is an interdisciplinary field that uses rational processes applicable to specific needs. It includes a division of labor, a hierarchy of expressed authority, and a framework of known rules and adaptive procedures. The industrial engineer, because of the wide scope of his possible work assignments and rotation through a variety of jobs that are key to company profitability, should be favored to advance in the ranks of management. He is also in a position to fully comprehend the need for human error prevention in order to achieve organizational goals.

Motivational concepts

It should be understood that some of the manager's basic motivational concepts have a long history. The French philosopher Rousseau, in 1762, indicated that organizations flourish only when individual and collective interests can be pursued to the fullest extent possible. It has been long known that people are motivated by work made interesting, by the opportunity to be heard and to contribute toward decision making and mutually beneficial goals, and to achieve appropriate personal recognition for good job performance.

The concepts of Maslow, McGregor, and the Hawthorne study previously discussed in this chapter are rather rudimentary motivational theories in terms of specific application to current sociotechnical complexities. The emphasis now may be on solving individual manifestations of employee maladaptive behavior to changed circumstances.

The manager should avoid layering (introducing more layers of bureaucracy) and establishing deliverables (artificial goals of numbers of items to show some sort of theoretical progress). These create inefficiencies and sour motivation. Creation of ever-more-centralized bureaucracies diminishes the role and possible motivators for individual achievement. The small (below 300 people) semiautonomous units help the perception of personal contribution and feelings of self-importance.

In terms of human error, a good manager should utilize motivational concepts as a first-line defense against human error. If there are residual problems in human system dynamics, the manager should call upon the resources offered by a relevant specialist.

Supplemental error principles

The industrial manager should be aware of the following old human error principles (as reported by Peters, 1963):

Adaptation of remedies

"People do not always act as they are supposed to, as they have been instructed or directed, or as you might suppose that they would or should. Hence, equipment and procedures have to be adapted (and readapted) to reduce or prevent the intrusion of unwanted human variability during periods of inadequate organizational control, work stress, and the typical operational handicaps which have to be expected on the basis of past experience."

Multiple actions

"There usually is no one simple solution to human error problems; it is more typical to find that there are multiple corrective actions required. In tracing back the causes which led up to or permitted the error to occur, it is more typical to find that there are a series of branching preventive measures which are desirable to help preclude the recurrence of such difficulty."

Error indices

"Avoid attempting to utilize figures-of-merit based upon human error to rate equipment items or attempting to use human error percentages as an index of workmanship, craftsmanship, or morale. Any comparative rating scheme will unfavorably influence opportunities to gain or evaluate information and will hinder negotiation for corrective action."

Arbitrary controls

"Intelligent workers can 'beat any system' of arbitrary or paperwork controls which might create some difficulty or unpleasantness for them or which seeks to impose upon them what they feel are unrealistic requirements."

Unreported incidents

"A flood of previously unreported incidents could result as people hear more about human error, become sensitive to it, and suffer no personal repercussions as a result of reporting it; this new information could easily change the ascribed causes of some chronically recurring problems."

Nuisance problems

"Do not ignore minor, trifling, nuisance, or marginal problems. They often serve as fruitful leads to significant problems."

Responsible supervision

"Superficial corrective action, such as the 'notification of responsible supervision' or 're-emphasis' of existing regulations, generally appears helpful in the absence of any other course of action. However, it should be recognized that habitual response of this type may be inadequate and that it sometimes does more harm than help."

Based on prior experience within a company, an industrial manager is often well aware of the probable cause of many human errors committed by his subordinates, himself, and other managers. Often the manager has formulated his own additional human error principles, whether effective or not. The principles above serve as reminders.

ERP teams

One of the most effective organizational safeguards against the proliferation of human error is the deployment of an error reduction program (ERP) team. This is particularly important for a project engineer or program manager who resides in one part of an organization (such as the engineering department), but his ultimate responsibility extends to where his designed product is being fabricated and tested. The ERP team can provide vital information as to what is really transpiring, forestall possible mistakes, and obtain quick action to correct a glitch in the process. This is particularly helpful for remote manufacturing locations and for critical suppliers.

The ERP team may be composed of the following job descriptors or dedicated functions. The *error detective* responds to error situations by directly investigating and consolidating the facts and circumstances. This investigation is direct and personal, not a simple reliance on the required written form reports of others or the bias that may be found in formal accident reconstruction conclusions. The error detective may also engage in proactive, broad net searches for causes of error and near-misses. There is usually an attempt to negotiate helpful corrective action at the worker–supervisor level. After the compilation of the investigative facts, the process proceeds to the next step. The *error integrator* reviews the compiled information in terms of oversight and possible needed action. He may consider the impact of the errors from the perspective of the project or program, the company policy and procedures, and the customer requirements. The integration is to determine plausible changes and what might occur to dependent variables. There should be some summary written record for the next step. The *error auditor* provides a time-delayed follow-up to ensure that preventive changes have been made, in what manner, and to what degree of effectiveness.

Those engaged in direct error reduction programs learn from experience and become sensitized to human error situations. In essence, they provide special training for future test observers or monitors. Without such training, there may be passivity and nonrecognition of important error issues during the observation of various tests conducted by others. Preparation is key to obtaining useful results from testing, and the special training on error reduction enables the test observer to better prepare for an active role during the test procedures. He or she has to know what to look for. Thus, the ERP teams can become a valuable tool for the project manager. Just the designation and activities of an ERP team send a strong message and should create a Hawthorne or placebo effect in terms of error reduction.

Cautions

The human error specialist may find errors emanating from sources outside his or her assigned responsibilities or jurisdiction. These sources may include conflicts of interest, financial mistakes, questionable accounting, and intentional circumventing of internal company or agency controls. These are matters generally handled by others in an organization; they may be troublesome quicksand, and they could create ethical dilemmas (see Chapter 12).

A somewhat similar situation may occur when there is an organizational crisis, transition, major new challenge, or temporary information overload. Significant errors may occur despite facile-appearing change management. The human error specialist should be particularly sensitive as to how such situations are handled in terms of diplomacy, jurisdiction, and effect. A welcome mat or a special authorization may be desirable. Derivative problems may be deemed beyond the competency, talent, or experience believed necessary for a skillful and productive engagement by the human error specialist. Beware of the unfamiliar, where decisions and errors are made under novel stress and errors may not be what they first appear. Tolerable or justifiable error may be acceptable and noncorrectable.

Caveats

Systematic error

An enterprise's overall mental health can be revealed and measured by its aberrant systemic human error symptomatology.

Emotional error

Since strong emotions can drive human error away from otherwise well-reasoned activities, it may be necessary to reshape images and de-energize any adverse perceptions of potential combatants by face-to-face friendship interactions. Friends are less likely than strangers to misinterpret company objectives and commit emotional human error.

Error detection

An affirmative effort to discover and correct significant *organizational human errors* is usually within the province of built-in organizational checks and balances, but they should be supplemented by periodic independent audits and technical assessment forecasts. Organizational errors generally seem rational and of merit at the time, so extrinsic evidence may be needed to support, help reaffirm, recast, or redirect an essentially unfortunate behavioral situation.

Cueing

Top-down management cues tend to gain strength and have implications attached as they are interpreted and communicated within an organized bureaucracy. To avoid miscues, errors, misdirection, and misguidance, and to maintain efficient targeted goal direction, some two-way transparency is necessary about the reasoning and desired end game.

Clarity

Leadership requires clarity as to the tasks to be performed, time schedules, appropriate enforcement, and recognition or rewards. Time pressures created by arbitrary schedules may serve to introduce opportunities for inadvertent human errors. Since rapid *error proliferation* can occur under the rush to complete a defined activity, timely procedural error correction procedures should be instituted. Obscuration leads to endemic error, so clarity in communication is essential.

Peer evaluations

Internal peer evaluations are important, but in-house evaluations may be biased, serve as part of a groupthink situation, or act as camouflage for inbred organizational error. For *opinion-based human error*, usually might makes right, except when there is the power or weight of external peer evaluations and audits.

Rigidity

An excessively rigid organizational structure may suffer brittle failure when subjected to technological, social, and competitive stresses. Command and control should provide for dissent, ample creativity, and the necessary adaptability to the unexpected.

chapter eleven

Management challenges

Introduction

There are some gross life-and-death policy issues now confronting most commercial and industrial enterprises and their requisite infrastructures. The timely implementation of appropriate and responsive policy decisions may be crippled by infestations of human error, or more hopefully, the corporate decisions may serve to vanquish most debilitating errors. The need for ongoing substantive policy changes has been induced by rapid technology advances, broad-scale information transfer, world trade interactions, global bureaucratic concerns, competitive costing, quality initiatives, and the social dynamics of the marketplace. This chapter, together with issues raised earlier in the book, should enable top managers to perceive and attempt to favorably resolve these challenges as they become manifest in varying forms.

Extended error traceability

An essential management control is the rapidly evolving track-and-trace systems that help eliminate human error. Historically, the concept of *configuration control* was needed to identify all of the components in a given product where each product could be different, as in aircraft systems. Parts replacement was predicated on the exact part used in an identified (serial numbered) product or system. Variations of this concept were used for ostensibly identical mass-produced products (using the batch or lot numbers of a production run, the manufacturing and expiration dates, and elemental bar code information).

The more modern concept is *traceability*, the track-and-trace historical record of a product. For example, by placing radio frequency identification tags on products, containers, cases, and pallets, it is comparatively easy to initiate and maintain life histories of each product on a computer system. Today, components from all over the world may be mixed, substituted based on cost and availability, or improved midstream in the production cycle. Precise historical information is needed from cradle to grave on most products or cradle to table for food products and their ingredients.

Traceability became more important because of some food poisoning cases, the advent of bovine spongiform encephalopathy (mad cow disease), and the fear of terrorism (how terrorists could affect the food supply). The European Union requires, by regulation, traceability of food and feed at all stages of production. In the U.S., there is a maintenance of records requirement for the food and beverage supply chain under the Bioterrorism Act.

There are evolving concerns about contamination and genetically engineered foods. Traceability and transparency are now essential for all products. The audit trail is also becoming an important management tool. It permits a contemporaneous monitoring and audit procedure that is becoming an important direct management tool.

In a manufacturing setting, human error can occur in almost any manual operation, such as data entry and product inspection. The conversion to automatic operations, by using bar code or radio frequency identification systems, has substantially reduced or even eliminated those sources of human error.

In a medical setting, patients may wear bar code bracelets. These are read by smart medicine-dispensing carts, so that the patient receives only the properly prescribed medicines. This reduces so-called medication errors and permits traceability by product code, lot number, and expiration date (Van Houten, 2005). Smart labels on drug packages permit automatic stocking and reordering. Clinicians use smart cards during stocking, and if incorrect drugs are placed on the cart, an audible warning sounds. Digital security is better ensured by electronic signature comparisons and smart card authentication. A printout given to each patient allows him or her to check that he or she has received the correct medications at the correct dose. Detection of drug counterfeiting, deterring diversion, and recording the chain of custody is achieved by random serialization in the bar code. Similar error reduction techniques have been used by pharmacies and supply management systems.

Establishing an effective traceability program is an important organizational responsibility. An inadequate traceability program may be recognized only after a costly product recall program has been instituted. The proof of innocence may be incomplete or missing. The speed of response may be so slow that the product brand and corporate image may be damaged. A fault in one item may be traced to other items. In other words, *traceability errors* become obvious when the data are finally put to some use, particularly in an emergency. Thus, the system should be exercised in periodic trials to uncover inadequacies, including human errors in the system that could be unfavorably interpreted by those in the marketplace.

Computerized automatic traceability systems may be further extended into the supply chain, inventory control, shipping, warranty, and customer complaints. They may be extended to the productivity and error score of each worker. With integrated computer systems, a central "war room" could be set up to permit contemporaneous monitoring of all events. This degree of visibility may help in understanding what is going on in the overall system and permit quick decisions. In some situations, such transparency may help to overcome an impermeable internal bureaucracy that justifies the status quo, rationalizes problems, opposes innovations, and cloaks errors. The disadvantages are too much detail to be digested, management control that may be too centralized and rigid, and the possible creation of a subservient worker attitude that could impede delegated local responsibility, quick improvisation, and hands-on problem solving. The advantages are an awareness of what is actually happening, quick identification of bottlenecks, and a better assessment of the overall health of the enterprise.

Depersonalization

The productive employee is generally a highly motivated individual who takes pride in his job. There may be some ego enhancement when it is revealed that he works for a company with a good reputation. Worker loyalty may be enhanced by incentives such as health care insurance and ample retirement benefits. Above all, the worker is a recognized person, a unique individual, and someone of special value. This type of recognition may occur in a small enterprise or a major conglomerate. It may be independent of wages, hours, and conditions of employment. The image is of a happy person and a good worker who is an asset to the company enterprise.

The same employee may be depersonalized by beliefs about social isolation, organizational remoteness, and interpersonal peer indifference. There may be feelings of being lost in a very big company, being only a face in a crowd, or a number rather than a person. Depersonalization may or may not be divorced from reality, but different realities are often perceived from one situation. In some companies there are constant redefinitions of performance targets, tight deadlines, and penalties of demotion or dismissal for failure to achieve the goals. Chasing an ever-changing target may provoke feelings of helplessness, lack of control, and machine-like demands. Failure and punishment lower self-esteem. Success has benefits, but no respite. Escape may be a possibility, but the seeds of depersonalization may have been already sown.

It has become a management challenge to provide a social situation at the work site that stresses the importance of each worker on a personal basis. This may be essential when there are reductions in retirement benefits, limitations on health care, job insecurity, factory relocations, mergers and acquisitions that dissolve corporate identities, and gross inequality in wages, salaries, and bonuses. Whenever there is fear and resentment, there may be an increase in depersonalization. If a worker does not care or is resentful, intentional human errors may increase. Depersonalization is associated with a lack of pride, inattentiveness, and uncorrected errors of omission or commission.

Workers exist in a social situation composed largely of work and commuting, home and family, friends and recreation, media and other sources of information, and other unique pursuits. These can either complicate or simplify management's attempts to acquire and retain good workers and take measures that could prevent depersonalization as a type of communicable disease. This management challenge is difficult because it is an unfamiliar role for tough-minded, hard-driven managers. It has become an important variable because of the consequential error probabilities, the cultural mix of worker pools, and enhanced notions of personal equality, equal treatment under the law, workplace health and safety, and social parity.

Incisive restoration

Many things can go wrong in the operation of a business. The challenges are to more incisively identify the causes of trouble, have a means to ensure a rapid response, create the ability to cultivate more effective long-term solutions, and implement procedures that will involve top management where short, fixed-time deadlines are violated and quick restoration of normal operations are jeopardized. Some of the techniques used may be the creation of properly trained "quality commandos" to conduct on-site joint audits in supplier plants where faulty parts have been produced. For in-house troubles, cross-functional teams could operate outside organizational fiefdoms. More collaborative agreements with suppliers (shared design) could result in early redesign actions rather than later field problem corrective action.

There are management attempts to move closer to the sources of trouble so that there can be more realistic, quicker, and incisive solutions that could result in the restoration of desired operations. There are cautions to be observed; for example, in determining who is responsible for a problem, the hardball blame game and cost allocation should be moderated, where there is good cooperation, reasonable compromises, and a significant ongoing relationship. Procedures that move an unresolved problem to the desk of a chief operating officer after 1 week or to the desk of the chief executive officer after 2 weeks are a very public way of demonstrating that top management is both concerned and involved. Appropriate problem solving becomes a reality that must be faced quickly. Problems cannot be covered over or ignored without peril if top management has a hands-on reputation.

When a product recall is deemed necessary, the focus of attention is generally on how to conduct the recall, in the least expensive and most favorable manner. Far less attention may be given to identifying the real underlying human error that caused the defect or resulted in the unacceptability of the product. The excuses pertaining to cause may be defensive in nature, as if no one in the company could have been at fault or really deserved the pointed finger of blame. The corrective action may sound good but be so superficial, vague, and ambiguous in character that nothing really changes. The rationalizations may produce an aura of innocent behavior. There may be an apology and an expressed willingness to be more careful. It is as if human error is something to be expected, an act of God, an excusable rare event, and that any change would be an admission of fault. The hope may be that everyone on "our side" will emerge unscathed. With this attitude, a problem may take on a life of its own.

There should be a recognition that almost all recalls, defects, deficiencies, rejects, complaints, and claims suggest human error by specific persons, at given locations, and at a particular time. If uncorrected, they will repeat the error. Intensive and incisive probing for the correctable cause of error may be uncomfortable, particularly among peers. Unfortunately, uncorrected problems often pass on to new products. In one case, the product analysis design errors continued to haunt three generations of a product over a 20-year span of time. In another case, the cause was ostensibly corrected only to reappear several times over several decades. It was unfortunate that the costs of each error were in the millions of dollars, with adverse public relations and significant market erosion.

The management challenge is to find real solutions to problems that in the past may have been overlooked, ignored, or subject to quick dismissal as relatively unimportant or not proven. Since the cost of customer dissatisfaction is escalating, the role and the function of corporate executives are being challenged. Being incisive (to the point) and attempting quick restoration (correction) of operations may not be easy using old conventional methods.

Outsourcing

There has been a major tide of outsourcing that has changed the character of most industrial enterprises. It might constitute a delegation of work to first-tier suppliers located in a just-in-time adjacent industrial park. It could entail independent domestic contractors or sources in remote countries. The work that is outsourced may involve product components, services, processes, and technology. It may start with the purchase of manufactured parts, but may gradually escalate to include design responsibilities. The reasons for outsourcing may be simple per part cost reductions, avoidance of business investment costs, enforceable cost controls, access to special technology, greater flexibility in fulfilling market demands, which may be boom or bust, and the management simplification of manufacturing only high-profit high-tech assemblies that are retained in the business model.

The response to rapidly changing technology, markets, or organizational needs may be the formation of joint enterprises, attempts to achieve sufficient stakeholder control of other related enterprises, or the forging of intellectual property agreements. When there is an occurrence of techno-shock, financial distress, dramatic change in market conditions, or unexpected and unfavorable customer perceptions of the company or its products, there may be an emphasis on short-term survival. This may encourage well meaning, but less advantageous business plans. Misguided management concepts and unrealistic goals may surface. The stage may be set for increased human error, which may accompany any transitional or change response. This is particularly true where there are employees who are resistant to change, who do not want to learn new skills or accept added responsibility, who may be habituated to or protected by contractual work rules, or are not capable of learning that which is expected and needed by the company.

When outsourcing does occur, questions may arise about software code glitches, cultural mind-sets, independent or self-managing conduct as contrasted with autocratic or need-to-be-told behavior, quality conformance decisions, the ability to deal with fuzzy data, and effective amiable oversight. More important are beliefs among company engineers that they are training engineers who may replace them or cost them their jobs in the future. They also fear that future jobs will be less paternalistic, with fewer perks and lower salaries, because they are training so many outside engineers and creating so many opportunities for competitive enterprises.

All skills must be continuously updated and upgraded or they will fade into a self-imposed mediocrity with attendant error. Techno-social change surrounds everyone, worker or manager, which requires refinement or adaptation of skill sets. However, all outsourcing transfers a potentially valuable opportunity to learn from an ongoing specialized business experience. It removes the opportunity for continuous skill enhancement. The company eventually may be deprived of a competitive skill bank. At the same time, the company receiving an outsourced contract can develop the skills of its workforce until it exceeds that of the original manufacturer. Outsourcing may develop superior skills in the supplier, despite attempts at retained control over general core specifications.

There is the belief and practice that training of company engineers is not really needed, that they can be hired and immediately be assigned to responsible jobs without much supervision. This may be to avoid dependence on any individual, whether entrance level or senior management. But this "no irreplaceable talent" required belief may combine with a "no opportunity to learn" situation. One response has been to hire many engineers in the hope that a few may spontaneously turn out to be what is needed. This may be a rather costly approach compared to the highly motivated, well-trained worker who is given the opportunity to perfect the real tailored skills needed by the company.

The management challenge is to retain sufficient opportunities for skill development that otherwise might be lost in outsourcing operations. It is not simply a make or buy type of decision, since there are core values in terms of experience acquisition, intellectual property in the form of patents and skills, and knowledge-based error reduction. Stockholder dividends may not be acceptable, over the long term, by the elimination of desirable features that may prevent customer injury, error, and dissatisfaction. The management challenges are multifold and complex, often governed by external forces and to some extent moderated by any expressed internal virus, the chief of which may be unrecognized and unabated human error. Outsourcing can be a brain drain, a quality system nightmare, and an error producer unless rigorously and appropriately managed.

Intelligence

The marketplace can undergo quick and dramatic changes, sometimes giving birth to surprisingly agile and viable competitors. This requires relevant, timely, and predictive knowledge (intelligence) about such changes in order for management to respond rapidly, appropriately, and decisively. There may be a good method to collect such intelligence, to analyze it, and to make recommendations. The most common errors at this point are misinterpretation, ignoring possible threats, separation of signals from noise, insensitivity, and ignoring recommendations as information is passed up and filtered within the chain of command.

The intelligence function might benefit from some of the skills, tools, and nimble creativity of the human error specialist, for example, separating actual causality from associations and coincidences or, in essence, the seeds from the chaff. One problem may be in determining the value of good focused experience in intelligence matters as balanced by the dulling effect of long-term socialization to given company norms, business customs, and a possible superiority bias that might be encouraged within a company. In other words, a personal belief that "we are the best" may blind or distort an assessment of facts about the marketplace. The company leader, usually the one person who really rules from the organizational roost, may have the insight, candor, knowledge, motivation, perception, and apprehension to realistically assess challenging information. But that kind of spirit may not be widespread within a company. In some companies, emerging managers are rotated around a variety of organizational positions to broaden their skill capability and to have persons that can do almost any job at a time of need. The depth of such job familiarization may not be sufficient in terms of intelligence collection and assessment, which has in-depth demands.

A company policy may indicate that it is a technology importer and expects synergies across product lines. Both impose change and heighten the probability of error. There should be an audit for correctable induced human error. The normal process of human error correction has a distinct analogy to the process of intelligence gathering, analysis, and transmission to affected persons within an organization.

Error R&D

A chief executive may want and encourage a high level of discipline within an organization. Discipline offers predictability of subordinates, a concerted unity of purpose, less time on conflict resolution, and simplicity in terms of decisions. Discipline generally means following company rules, whether formal or informal, oral or written policy, or just something inferred by custom and practice. Breaking the rules is a significant error. It may adversely effect company operations to some degree. Some errors may have a latent effect in terms of discovery or consequences.

In general, violations of discipline are unacceptable. There may be some allowance for desirable creativity or independent action on nonroutine fix-it matters. Too much management leniency may induce uncontrolled Darwinism, with survival of the fittest attitudes, open rivalries, and further disruptive disobedience. There can be healthy competitive behavior, channeled within the rules of good corporate behavior, to encourage goal-directed and highly motivated activity. In fact, selection of managers may be primarily based on aggressive, high-energy, can-do, self-starter behavior. This may set the stage for less cautious acts, greater discipline errors, and governance problems.

The human error specialist may not want to enter the intermanager problem-solving arena, the corporate misjudgment scene, or the playing field of human resources or recruiter specialists. But need may create a demand for the special skills of the human error specialist. Assuming that there are unique error problems and an urgent need for creative solutions, the time may be ripe for a breakthrough by some error R&D. Just as technical R&D is funded, the error problem can be treated in a similar fashion.

It does no harm to openly discuss human errors, formulate hypotheses, perform critical observations and experiments, eliminate erroneous concepts, and formulate theories from sets of hypotheses that have been confirmed by empirical evidence. This is far better than the hunches and guesses of the nonspecialist. Experience-based beliefs may or may not be valid. If they do not work, perhaps special action is justified by trained specialists in error reduction.

The behavioral aspects of empathy and social responsibility are illustrative. A manager may be impaired in the ability to empathize and understand the behavior of others. This may be related to decoding the intentions, emotions, and future actions of others, a desirable attribute for managers of people and projects. The question may be whether there is weak brain activity in Broca's area, a miscue in the brain development of the right frontal cortex, or a loss of predictive ability in the parietal area. This suggests that the error-solving approach should be grounded in scientific findings, tested in similar environments, and capable of yielding results that have practical application with beneficial results to the organization.

Contagion

The chief executive office of a corporation assumes a leadership role, and his subordinates are expected to follow the leader, implement his policies, and echo his pronouncements. In fact, all company officers are vested with some authority, power, and a sharing of the father figure image. It is not surprising that other employees will reflect the personal attitudes that are expressed in or inferred from decisions, statements, memoranda, appearances, and the demeanor of company officials. Bad attitudes can percolate down the organizational structure and spread within the entire company. A negative attitude may envelop the entire enterprise. It could become a contagion affecting supplier relationships. Negativism may affect customer perceptions from personal interactions, such as how warranty claims are interpreted, processed, and resolved. It may become an aspect of company culture that displeases various government officials, from regulators to tax assessors. It may erode the prospects of joint ventures with other companies. It should not be dismissed as simply a minor factor compared to the money motivation.

Negative attitudes are an important factor in error causation. They may serve to relegate error correction and prevention to an unnecessary status. If negative outcomes are presumed, the status quo becomes the defensible objective within company functions. Nothing really attempted means no lost battles in the power struggles that serve to energize companies, and there may be gradual disillusionment among progressive contributors to the company fortunes. There may be slow erosion of company market share, and motives then emerge for self-aggrandizement during the decline of the enterprise, beyond turnarounds, reorganizations, and growth by acquisitions.

A radiating positive attitude combined with diplomatic prowess, displayed and reputed to top management, signals a far different scenario and company prospect. Employees, suppliers, and customers respond favorably to positive, friendly, welcoming, and appreciative attitudes. Creativity flourishes where it is welcomed and rewarded, whether it is creative product design or creative marketing. Positive motivation means that employees are convinced that company objectives will be achieved on time and within cost estimates. A positive and understanding culture is one in which human errors are revealed, not covered, and error countermeasures are real, not illusionary.

The challenge to management may be to feed the contagion of positive attitudes that are oriented toward achievable enterprise goals. This can be helped by strong error correction efforts by competent specialists. While employees' pride about a company, as a whole, should be encouraged, there are examples of companies that have exuded attitudes of expressed superiority that was intolerable to others. In fact, that kind of superiority may be a manifestation of negative attitudes rather than behavior associated with more acceptable corporate attitudes and culture. This suggests that positive attitude development is not a superficial quest.

Cognivitis and linguistic dystrophy

A chief executive may be exposed to an excessively large volume of information, and much of it may turn out to be useless clutter or be contorted by layers of based selective filtering. There may be information suppression by management teams who know what the boss wants to hear. There may be covert self-promotion in the relative importance of what is presented. There may be information from many projects, groups, and initiatives. There may be gaps that contribute to ambiguity and uncertainty. There may be conflicts in the information available for decision making or for the informed monitoring of operations. The information overload may produce cognitive confusion (cognivitis) and a reduced ability to communicate (linguistic dystrophy). The result may be errors of assessment and judgment.

The executive may believe he is too close to all the details and backs off for some peace of mind. This drift away from cognivitis may result in ignoring what could be important information. Indifference could breed a malaise and produce maladroit decisions. Indecision could increase infighting among subordinates and the executive could be blind-sided. Less and less may be known about the thinking on the shop floor and the difficult problems in supply chain management, enterprise resource management, and harmonizing the diverse operations of the enterprise. Linguistic dystrophy may be reflected in meetings of the board of directors, stockholder meetings, and media commentaries.

The top company officers may become remote in order to strategize about future directions and opportunities for the company, leaving the day-to-day operations to others. Some remote or complacent executives have been greatly surprised by the turn of events in the marketplace or in the fighting between autonomous business units. The cost-effectiveness and monetary return of various initiatives may become obscured. Corporate frugality, cost cutting, and scrap reduction efforts may lapse. Organizational disarray and dysfunction may appear. There may be problems in flexible manufacturing and build-to-order programs, value pricing, product development schedules, recall and warranty cost reductions, modular outsourcing, local content, and customer satisfaction indices. Business plans may become rigid or ignored. Pain sharing ensues. Knee-jerk decisions, under pressure, may prevail, and obvious error abounds.

The management challenge is to find the right balance between an excessive indigestible information overload and remote governance to minimize costly mistakes and obvious errors. It includes how to bring order and direction to a company during normal transitions and change, ensuring cross-company cooperation, and maintaining a competitive culture. It should include a means of experience retention so that conflict resolution serves as an example to those who might repeat violations of prescribed company culture. What is the proper balance between reading executive summaries as contrasted to personal observations and practical knowledge of assembly lines and their production problems? Most important is a means of integrating a company consisting of employees, shareholders, suppliers, labor, and customers.

Rising stars

It is an important management challenge to identify the rising stars within the organization. It is well understood that they could be the future lifeblood of the company. They should provide energy in an appropriate style, without weaving or jumping around issues in a manner that creates organizational inhomogeneities. They should be able to think, talk, and operate in a logic-built manner, act within relevant frames of reference, be willing to disclose problems, and be able to undertake quick remedies. This means openness, transparency, and understanding the reasons for due diligence. Counterintuitive options should not be cavalierly rejected without well-reasoned logic and foundational facts, since the enterprise's competition may quickly reject an option and thereby overlook a possible opportunity, or vice versa.

The desirable personal attributes of rising stars could serve both the error correction (post event) and error prevention (future event) efforts.

If assigned to work with error specialists on just a few cases, they could enrich the effort while learning the basic techniques. When they move ahead in the organization, they will have a better awareness of human error, its causes and control, and can take immediate action with or without the help of error specialists. Most probably, the human error team efforts would sensitize the rising stars so they would know when to call for help and have the know-how to avoid the possible ineptitude and impotence of less sophisticated reactions to error.

Robotization

There is a mantra that there will be an increasing number of robots in future factory assembly operations and that this type of automation will reduce the number of workers required. It also could reduce human errors because the machines are more consistent in what they do. The robots can be introduced workstation by workstation, as an incremental process in factory modernization. The robots range from very simple devices to programmable multifunctional devices useful in flexible manufacturing. The robots might be smart or self-adjusting, so that there is not a need for so many parts decisions by the worker on a flexible assembly line producing different or customized products. This may be combined with kitting, where off-the-shelf parts in storage are placed in packages and delivered in synchronized fashion with the flow of units on the assembly line, so the assembly worker has what is needed when it is needed. The objective is to reduce complexity and decision making for the worker, which means simplification of product systems and elimination of errors.

Industrial robots can lift heavy loads, carry them, hold them while welding occurs, and then pass them on. They may reduce the need for material conveyors, equipment to hold the parts during work on them, and worker involvement in transfer operations. One worker may monitor several robot workstations.

The management challenge is that the use of robots is costly and time-consuming when first introduced. Lengthy experience is often needed when robots are used for mass production or complex manufacturing technology. The challenge is that labor costs per unit must be reduced in a competitive marketplace, and the consequences of error must be minimized to ensure profitability. So exactly what type of robots are needed at what time frame and factory location? It is an investment in a rapidly changing technology. Errors that do occur with robotization may be more persistent and difficult to correct. The error remedies involve stabilization of the output of precision robots requiring different human skill levels that may or may not be available.

Error detectives

There may be a significant management challenge if the error detection efforts are authorized to cross-organizational borders (company hierarchies). What skills should an error detective team possess? What is the appropriate mix of personal experience, the authorized procedures used, objectives, and targets? Based on the experiences of the authors of this book, only about one third of such efforts were sufficiently productive, diplomatic, and served to actually advance company goals. A two-man team may be advisable, consisting of a creative and incisive outsider (new eyes) and a staff expediter type (familiarity with how the company operates). This has proven effective in large manufacturing, test, and logistics operations. The authors have found that in relatively small operations, independently managed, a one-man error specialist can operate informally, use backdoor access and direct observation, and personally advise immediate management as to what is actually happening on the factory floor.

The error specialist may be in a good position to advise on lean operations and direct cost cutting with an outside perspective. This may be far more appropriate than a very general across-the-board 10% cost reduction (budget allocation) or reliance on normal attrition losses. For example, a holding company mandated a 5% cutback for each of 5 years on an independent subsidiary. Starting with the second year, the cost savings were focused on cheaper material alternatives and sourcing. The resulting product quality problems began to appear 1 to 2 years later and totally overwhelmed the cost savings that were achieved.

Any attempt at intelligent downsizing and targeted cost savings by use of supplemental information from error detectives should not be at the sacrifice of their primary function. Indeed, error reduction does directly result in cost reductions as well as eventual customer satisfaction. Ambiguities, generalities, and techno-babble are no substitutes for independent hands-on knowledge acquisition and evaluation.

Caveats

Brain drain

Outsourcing may provide immediate cost savings and other benefits, but can gradually degrade a company's skill base, error control, and proprietary technology assets if not properly managed.

Visibility

Expanded traceability procedures can facilitate direct management control of products and processes, could provide error scores for each worker, and can complement any search and produce requirements for documents, communications, events, decisions, and accountability.

Loyalty

An industrial enterprise is a collection of unique individuals, each with varied skill sets and talents, organized and motivated to achieve common goals at specified times. Individual productivity, company loyalty, and error rates can be dramatically altered by management actions that affect social and personal values related to chronic depersonalization.

Remote control

The function of top management is changing from remote control to hands-on knowledge that permits incisive problem solving and quicker restoration or adaptation of operations. The demands of a large corporation may relegate this to a selective sampling of what might be occurring or resulting from corporate inbreeding.

Direct feedback

There should be some direct feedback loops between top management and the customer/user base, such as clearly established Internet forums for user comments and complaints. The feedback loop should be intended to minimize possible middleman interception, masking, distortion, blocking, disregard, or excessive delay. If an open (easily tapped) feedback system is utilized, it could be a means for early unfettered discovery and correction of latent or unrecognized error problems by the human error specialist.

Leadership

There is much to be learned about human behaviors in industrial and institutional social settings. Rational choice behavior, predicated on only self-interest, has been challenged by recent scientific findings. Subgroups form alliances, develop loyalty subsets, and have limited communication with those having different traditions. Dynamic social settings create changes in social circumstances that can markedly affect common causes, loyalties, job performance, and the occurrence of significant error. The management leader must respond in a timely manner in terms of appropriate institutional structure, fair and sustainable rules, and targeted initiatives that are relevant to new competitive objectives, and develop a keen understanding of the behaviors, personalities, and propensity for human error that he must successfully manage.

chapter twelve

Professional responsibility (ethics)

Obligations and duties

Human error and safety specialists assume a rather special duty to exercise something more than ordinary care in the performance of work tasks and the expression of opinions. The duty is a moral obligation that is owed to others who might be harmed by the actions or inactions of the expert specialist. This is an obligation to the general public, particularly those in direct contact with a particular product, process, or system. The duty to peform in a competent manner is also an obligation to the immediate client, employer, and sometimes to a regulatory government agency. The duty may be codified in ethics rules, principles of professional responsibility, codes of ethics, the common law, or legislative statutes. Some professional societies may have watered-down generalizations concerning professional conduct, so that almost any behavior is permitted. Other professional associations may have detailed codes of ethics, interesting interpretive decisions, and strict enforcement activities (ECPD, 1974; NSPE, 1976; HFES, 2004). The competent professional may impose upon himself an even higher level in the performance of perceived duties and has a good reason to do so.

Those performing human error and safety tasks often hold in their hands the future well-being of many other people. As a result of how those tasks are performed, there may be resultant accidents, incidents, failures, or malfunctions. There may be consequential physical injuries, psychological trauma, financial distress, corporate economic problems, damaged reputations, and company marketing difficulties. However, the consequences may become manifest remotely in time and place, so the specialist may not be confronted with accidents in the near term and personally face those who become injured or otherwise harmed.

The remote nature of the harm serves to highlight the need for more immediate moral and ethical guidelines. Generalizations to do good just may not help. Rules of ethics are based on past history and real needs, so responsible persons can do better in the future.

Useful principles and practice

The following commonsense caveats and formulations of ethical principles may reinforce some existing codes of ethics, fill in some voids, assist in difficult decision making, and promote a better understanding of what is actually meant by professional responsibility:

1. *Avoid blind trust*: Do not assume that someone else is performing part of your duties. They may not be. In fact, they may not know how to do what you want or expect. They may be relying on you, since it may be your prime responsibility. Know the limits and character of what is expected from you. When human safety is involved, avoid a blind trust of others. This includes caution when utilizing data gathered or information compiled by others. The data may have been accumulated and processed for some other purpose. It may have been filtered, biased, or interpreted in a manner to avoid safety, health, or environmental implications. In fact, there might have been a fear of assessing or reporting a customer complaint as including a potential human safety problem. By asking critical questions, the specialist may determine the purpose, practice, and utility of warranty data, field trouble reports, adjustment returns, and customer complaints. It may be important to differentiate customer satisfaction, goodwill, or public relations objectives that affect the data-gathering process. Blind trust may be misleading, but informed trust may serve to reveal and yield significant safety information.

2. *Retain independent professional judgment*: Professional integrity is priceless. Without it, credibility is lost. Others should not want to rely on the quicksand of unpredictable and easily changeable judgments or opinions. Similarly, foundation matters. If the blind trust syndrome is not avoided, human error control may be unreasonably placed in the hands of others not actually qualified. Uninformed decisions by the unqualified may lead to mistakes, misinterpretations, and errors. Qualification means the possession of basic relevant sophisticated knowledge about human behavior, human error control effectiveness, legal implications concerning human conduct, and symptoms of human disease impairment, together with an appreciation of individual differences, cultural effects, and perceptions combined with appropriate scientifically sound principles, acceptable engineering analysis, and reasonable logic. In other words, there should be an informed basis for judgment that is beyond common knowledge. Most people believe they know a great deal about human behavior gained during a lifetime of living in a complex human society, but usually this is confounded guessing. Others do possess some useful knowledge. Only a few are real

experts in the professional sense whose opinions do have great value, credibility, and precision. The image and reality of independent professional judgment can be retained by actions that are courteous, cooperative, and agreeable to others who may believe or voice other opinions. Personal character and demeanor do count.

3. *Be proactive*: Conformance to legal criteria may be misleading, since criminal law cannot be applied *retroactively*, but the civil law that is in existence at the time of an accident, trial, or adjudication is effective as of that *future date*. In other words, the design application may be this year, but the design can be judged by what the law might be in 10 or 20 years. Of course, in the intervening time there is the opportunity, and in some countries the duty, to modernize or update, recall or replace, service or repair, buy back or remanufacture, and provide relevant advisory or warning information. Thus, a proactive perspective is necessary in terms of the shelf life, service life, storage life, and useful product life. Similarly, trade standards are constantly revised and expanded in coverage, although compliance as of the date of design, manufacture, or installation may be a good legal defense or excuse. There may be future superstandards, such as those on social responsibility or those that incorporate by reference a broad array of related standards, specifications, or guidance documents. Again, a proactive demeanor is required to personally investigate, gather, interpret, and anticipate what may be needed for the future use or operation of products, processes, or systems.

The Nuremberg Code

Shortly after the conclusion of World War II, the Nuremberg Military Tribunals were convened to determine possible criminal culpability and punishment. Out of this activity, 10 clauses or canons emerged concerning research experimentation using human subjects (*Trials*, 1947). The Nuremberg Code reflected certain basic principles pertaining to the acceptable moral, ethical, and legal aspects of such research.

Those who now study human error have a preference for data so closely connected to humans that it may be considered human experimentation. Data from animal studies and uncontrolled human research data are subordinated or suspect in terms of immediate translation to real human behavior in complex social situations. Thus, the Nuremberg Code has wide applicability in modern scientific endeavors.

The following canons were derived from principles from "civilized peoples," "humanity," and the "public conscience." Remember that the Nuremberg Code has become a cornerstone of professional ethics and has wide international acceptance.

1. *Free choice and informed consent*: "The voluntary consent of the human subject is absolutely essential. This means that the person involved should have legal capacity to give consent; should be so situated as to be able to exercise free power of choice, without the intervention of any element of force, fraud, deceit, duress, over-reaching, or other ulterior form of constraint or coercion; and should have sufficient knowledge and comprehension of the elements of the subject matter involved as to enable him to make an understanding and enlightened decision. This latter element requires that before the acceptance of an affirmative decision by the experimental subject there should be made known to him the nature, duration, and purpose of the experiment; the method and means by which it is to be conducted; all inconveniences and hazards reasonably to be expected, and the effects upon his health or person which may possibly come from his participation in the experiment. The duty and responsibility for ascertaining the quality of the consent rests upon each individual who initiates, directs or engages in the experiment. It is a personal duty and responsibility which may not be delegated to another with impunity."
2. *Beneficial result*: "The experiment should be such as to yield fruitful results for the good of society, unprocurable by other methods or means of study, and not random and unnecessary in nature."
3. *Justification*: "The experiment should be so designed and based on the results of animal experimentation and a knowledge of the natural history of the disease or other problem under study that the anticipated results will justify the performance of the experiment."
4. *Avoid unnecessary injury*: "The experiment should be so conducted as to avoid all unnecessary physical and mental suffering and injury."
5. *Possibility of death*: "No experiment should be conducted where there is an *a priori* reason to believe that death or disabling injury will occur; except, perhaps, in those experiments where the experimental physicians also serve as subjects."
6. *Risk compared to importance*: "The degree of risk to be taken should never exceed that determined by the humanitarian importance of the problem to be solved by the experiment."
7. *Adequate preparation*: "Proper preparations should be made and adequate facilities provided to protect the experimental subject against even remote possibilities of injury, disability, or death."
8. *Skill and care*: "The experiment should be conducted only by scientifically qualified persons. The highest degree of skill and care should be required through all stages of the experiment of those who conduct or engage in the experiment."
9. *Withdrawal by the subject*: "During the course of the experiment the human subject should be at liberty to bring the experiment to an end if he has reached the physical or mental state where continuation of the experiment seems to him to be impossible."

10. *Termination by the scientist*: "During the course of the experiment the scientist in charge must be prepared to terminate the experiment at any stage, if he has probable cause to believe, in the exercise of good faith, superior skill and careful judgment required of him that a continuation of the experiment is likely to result in injury, disability, or death to the experimental subject."

Note: The informed consent provisions of canon 1 have had great effect on the development of warnings designed to prevent human error. Frivolous experiments with "volunteers" have been reduced by compliance with the benefit–risk considerations implied in canons 2 and 6. The lack of follow-up investigations, diagnoses, appropriate care, and necessary treatment of human subjects, who have been harmed by experimental research, has been reduced by the inferences from the canons that there could be acquired latent injury or the delayed manifestation of injury. The use of children and incompetent adults in experimental research seems to be prohibited by canon 1, but such research has been conducted under circumstances of peer board approval, strict surveillance, and long-term follow-up evaluations.

In conclusion, the Nuremberg War Trials served to internationalize the concept that each person must consent to his or her personal exposure to risk, and that such consent must be informed, voluntary, and revocable.

Forensic principles

Those involved in human error activities should expect to become involved in the legal process sometime in the future. It may be to help answer written interrogatories (questions), to help find or provide documents that might be responsive to discovery demands (requests for production), to inform or assist legal counsel in technical or scientific matters, to testify as a percipient or expert witness, or to become a plaintiff or defendant in a workers' compensation, common law, or some regulatory action. In terms of personal professional responsibilities, there are some common forensic principles grounded in terms of basic concepts of ethical behavior. Some operative guidelines are as follows:

Independence

All testimony should be fact based, whether for investigative statements, declarations or affidavits, depositions, or in court. It is unethical to assume the role of a legal advocate, to speculate, or to rely upon dubious, misleading, or questionable information. Testimony should be perceived as credible, unbiased, and based on objective information. The capability to interpret data, make inferences, and arrive at independent conclusions is reserved for the court-qualified expert witness. The most important aspect of independent testimony is the comparative freedom from bias and direct truthful answers that create the impression of a believable, credible, and honest witness.

Confidentiality

In a professional relationship, the secrets of the client should be preserved and defended. This duty to preserve confidentiality pertains to trade secrets, protective orders, nondisclosure agreements, proprietary information, documents marked confidential and produced in a lawsuit, and information provided to an expert witness by a lawyer during the progress of a lawsuit. It is unethical to reveal proprietary information without the informed consent of the client, unless necessary to prevent a criminal act likely to result in substantial bodily injury or death. Generally, the client must be informed of the decision to release damaging information, and only after a good-faith effort to persuade the client to discontinue a course of criminal or illegal conduct. Disclosure is a last resort to prevent prospective harm to human life. Each state may have a public policy, defined by code or case law, regarding the protection of confidential business information. Before any release of such information a lawyer should be consulted as to possible exceptions to the duty of confidentiality, since there may be limited exceptions. In general, there is no affirmative obligation to reveal information to prevent harm. However, there is an ethical duty to disclose nonconfidential information when asked. An expert witness should not conceal, give half truths, mislead by the choice of technically complex words or concepts, or attempt to mask by purposeful evasion.

Integrity of others

The rules of the game do not embrace impugning the honesty, derogating the competency, or attacking the integrity of opposing expert witnesses or consultants. Belittling is quite different from disagreeing based on credible data, engineering logic, or scientific principles. Appropriate professional demeanor, basic civility, and reasonableness under the circumstances should be an important consideration. Differences in opinion may be based on different sets of facts, personal perspectives, or beliefs and values shared by peers in another discipline. For example, the occupational safety specialist may quickly fault the injured worker (it is his human error), whereas the design safety specialist may perceive the fault to be defective equipment (it is machine-induced human error). Of course, there are rare instances of witnesses who intentionally misread data or deliberately interpret wrongly, where professional licensing rules of ethics require reporting the wrongdoing. However, there may be other ways to correct the ethics problem. Thoughtful consideration and respect for the integrity of others may serve to enhance the reputation of the proponent, the image of his professional specialty, and attitudes about his ultimate client.

Fees

The fees and chargeable costs for the professional services rendered should be reasonable under the circumstances. It is unethical to charge unconscionable fees, unearned fees, or to accept substantial gifts from clients. The amount of the fees should be in proportion to the time spent, the novelty of the services performed, and the level of difficulty of the tasks required. There should be no fees on a contingency basis. There should not be circumstances that might create perceptions of purchased testimony or hired gun advocacy. Extra fees for employees involved in litigation are impermissible. In advance, there should be a mutually agreed upon fee schedule, specified tasks to be performed, estimated time and dates for performance, and a clearly defined role for the person in the legal process.

Familiarity with the law

The human error specialist should be familiar, in general, with the legal requirements that pertain to his professional services. For example, the term *foreseeable misuse* has both a legal and a design meaning. It serves to help define the legal requirements for the prevention of human error. The prospective expert witness should become familiar with his anticipated role in possible lawsuits and should inquire from attorneys about the basic applicable law in various jurisdictions that might pertain to his professional services.

Records

Carefully segregate and maintain business records in well-organized files. For the expert witness, this involves a list of each litigated case, the caption, the parties involved, its resolution, the time and fees involved, and his role in the case. In addition, qualifications such as academic achievements, employment history, publications, awards and honors, patents, and memberships in professional societies should be available. Such records may be the subject of questioning or requests for production in a lawsuit, but discussions of the content of ongoing lawsuits should be limited to that approved by the attorney in each case. The records should not be used as a basis for public statements. When a new case is accepted by the expert witness, an inquiry should be made as to what records should be maintained, what is required and can be disclosed legally, and the eventual disposal of the records. Consulting old records can help repair a series of "I don't remember" responses to questioners. Failure to maintain appropriate records might be considered an ethical breach by opposing counsel in a lawsuit, by a potential client, or by those attempting to determine possible conflicts of interest (Peters and Peters, 2005).

Other relationships

An expert witness cannot retain his impartiality if he is aware of an opposing counsel's personal or business relationship with the expert's family members or business associates. Such relationships should be disclosed in writing to the client (lawyer).

Client property

When employment in a civil action is terminated, all client property should be promptly released to the client (lawyer). This includes correspondence, transcripts, pleadings, exhibits, physical evidence, expert reports, and other material reasonably necessary to avoid prejudice in continuance of the legal process, whether or not paid for by a client. Any part of an advance fee, not earned, should be refunded.

Practice of law

The human error specialist should not engage in the unauthorized practice of law or help others to do so. It is unethical and illegal to directly or indirectly share fees with a licensed lawyer.

Collegiality

Some expert witnesses believe that there should be collegiality between opposing experts, but many trial lawyers are uncomfortable with this approach in highly contested cases. One fear is inadvertent or inferential disclosure of facts, theories, or tactics.

Institutional review boards

In performing experimental research on humans, there may be some unknown, poorly defined, or underestimated conditions that could cause harm to a particular human subject psychologically or physically. In the past, some subjects have been harmed by well-intentioned and respected scientists. To prevent such harm, there should be an independent review of proposed human research protocols to ensure adherence to accepted ethical principles as well as to methodological correctness and possible beneficial outcomes. In various countries, this assurance has resulted from the activities of independent research ethics boards or institutional review boards.

In the early 1950s, penicillin was available for the treatment of syphilis. In what has become known as the Tuskegee case, some men who participated in a syphilis study were not treated but merely observed. The New Orleans case involved observations of workers exposed to different exposure levels of airborne asbestos who were not warned, but merely observed. The question arises: Was scientific observation for a good societal purpose more important than the welfare of a few human subjects?

In 1962, the thalidomide calamity generated considerable public interest in the ethics of human experimentation. The World Medical Society adopted what has been called the First Declaration of Helsinki, in 1964, and the Second Declaration of Helsinki, in 1975 (Declaration, 1996; World Medical Association, 1964 to 2004). Researchers had been running experiments with little external constraints, recruiting research subjects by offering money or college credit, interpreting the Nuremberg Code as simply a discretionary guide for others, and there were many allegations of the unethical abuse of human subjects. It was recognized that "progress is based on research which ultimately must rest in part on experimentation involving human subjects" and that "the well-being of the human subject should take precedence over the interests of science and society" (World Medical Association, 1964, items 4 and 5).

One of the more important conclusions was that:

> The design and performance of each experimental procedure involving human subjects should be clearly formulated in an experimental protocol. This protocol should be submitted for consideration, comment, guidance, and where appropriate, approval to a specially appointed ethical review committee, which must be independent of the investigator, the sponsor or any other kind of undue influence. This independent committee should be in conformity with the laws and regulations of the country in which the research experiment is performed. The committee has the right to monitor ongoing trials. The researcher has the obligation to provide monitoring information to the committee, especially any serious adverse events. The researcher should also submit to the committee, for other review, information regarding funding, sponsors, institutional affiliations, and other potential conflicts of interest and incentives for subjects. (World Medical Association, 1964, principle 13)

There was increasing concern about the disparity between actual practice and the ethical guidelines for human experimentation. Voluntary compliance was a failure and formal ethics review boards were believed necessary (CIOMS, 1991, 1993; Council of Europe, 1990, 1996; Belmont Report, 1979; Levine, 1986; International Conference on Harmonization, 1997). Finally, institutional review boards were mandated, in the U.S., to review all federally funded human research under the provisions of the National Research Act (1974), the Protection of Human Subjects (21 CFR 50), and the Institutional Review Board (21 CFR 56). In Europe, there was the Good Clinical Practice for Trials on Medicinal Products in the European Community (1990), the International Conference on Harmonization's Harmonized Tripartite Guideline on Good Clinical Practice (1996), and the EU Clinical Trials Directive (approved in December 2000 by the European Parliament). The trend is to make ethical principles based on the Declaration of Helsinki into enforceable law.

Protocols

Formal written research protocols are used to secure preliminary human research agreements, to obtain subsequent ethics committee approvals, to ensure conformance during the actual research project implementation, to provide for informed audits during the research endeavors, and to constitute a permanent record of the research.

They may include:

1. *General information*: The protocol title, identification number, date, sponsors, monitors, investigators, and location.
2. *Objectives*: The protocol should include a statement of the purpose of the research. Why is the research being conducted?
3. *Subjects*: The research protocol should include an explanation of how and why the subjects are to be selected. It should include invitations to participate, inclusion and exclusion criteria, and inducements or expenses given to subjects.
4. *Procedures*: The research protocol should include a detailed description of the procedure to be used. This includes whether blinded, the statistical design, the placebos, and randomization procedures. The randomization code should not be unbinded, but provide the reasons and procedures should it occur. It should include the types, frequency, and duration of testing. It should identify the procedues for accounting for deviation from the research plan, including missing, spurious, or rejected data. An explanation of the procedures should be available in a separate referenced document.
5. *Risks*: The protocol should provide a description of all reasonably foreseeable risks, the functions of a safety monitoring committee, and the procedures that help avoid exposing subjects to unnecessary risks, inconveniences, and discomforts.
6. *Benefits*: The protocol should include a description of all benefits that might result from the research.
7. *Alternatives*: The protocol should include a disclosure of all appropriate alternatives to the research. Is this research really necessary, is it just a duplication of other research, or does it have some independent significance?
8. *Confidentiality*: A protocol should include a statement as to how confidentiality and privacy will be protected before, during, and after the research.
9. *Compensation*: The protocol should include a disclosure of any compensation for participation in the study by the subjects.
10. *Contacts*: The protocol should identify the names and telephone numbers of persons who may be contacted to answer questions about the research, possible injuries therefrom, and the rights of the subjects. The contacts should have special competency in both ethics and research procedures.
11. *Injury policy*: The protocol should include the institutional policy on research-related injuries and the availability of both diagnosis and treatment for adverse effects.
12. *Withdrawal and termination*: The protocol should include a statement that each subject may withdraw or refuse participation at any time,

and in addition that the participation can be terminated by the investigator without the subject's consent. If there should be replacements, include how they would be selected and included in the research design. The right to discharge subjects should not be used as an inducement or punishment by the investigator.

13. *Vulnerable subjects*: The protocol should include a description of special problems that could arise from physical disabilities, cognitive impairments, economic disadvantage, coercion, or undue influence.

14. *Informed consent*: The protocol should include a copy of an approved informed consent form and indicate that this form, when signed, will be given to each subject or the subject's legally acceptable representative. The form should contain language that the subject does not waive any legal rights or release anyone from liability for negligence. It should be in a nontechnical form and indicate the opportunities to inquire about the details of the experiment. The informed consent provisions are the most scrutinized in human research evaluations.

15. *Contract research*: The protocol should include an identification of all outsourcing and contract research, why it was necessary, the qualifications of the principals and investigators, its cost, and its adherence to the research protocol.

16. *Unforeseeable risks*: The protocol should include a statement as to whether the research procedure may present risks that are unforeseeable and a description of the follow-up of subjects who experience an adverse consequence from such events.

17. *Additional documentation*: The protocol should include provisions for the retention of all signed agreements, notices, warnings, forms, instructions, and relevant correspondence.

18. *Prior research*: The protocol should include a summary of all prior research having foundational merit, relevance, or instructive benefits.

19. *Final report*: The protocol should enumerate all reports required during and upon completion of the research. Generally, it would include a summary of the research outcome, and the publication policy if there might be publication in the scientific literature.

20. *Financing and insurance*: The protocol should disclose who is financing the research and the insurance coverage, its limits, the dates, and the named insureds.

Supplements

In some situations, the following may be required:

1. *Surveillance*: The research protocol should include a statement that persons or subjects will not be an object of surveillance technologies that would result in the violation of any of their rights, particularly an invasion of privacy. This may include profiling people's habits or behavior, that is, behaviorial research. An example may be the market

research endeavors where shoppers are videotaped without their knowledge. Secret digital videotaping may monitor individual or group behavior, eye movements, facial expressions, payments, distractions, demographics, and reactions to merchandise displays. Privacy may be maintained by computer processing and subsequent destruction of the images. Similarly, millions of cameras monitor people, for security purposes, in banks, hotels, restaurants, schools, offices, casinos, airports, and on the public streets. The digital universe is huge and fed by tens of millions of cameras. Is a simple sign that videotaping is in progress a sufficient warning or release, when people are already in the area when they have to interpret the message on the sign? This illustrates the ethical problems involved when surveillance is part of a research project. Surveillance may be desirable as evidence that the subject's rights were not violated.

2. *Information security*: The research protocol should include a statement that personal data will be protected as confidential, with the person or subject having the right to access, correct, or delete the data.

3. *Personal integrity*: The research protocol should include a statement that there will be due respect for each subject's physical and mental integrity and his or her dignity and freedom.

Caution: There is considerable debate about the need for so much detail in the research protocols and their application to so-called harmless research, self-regulated research, or minor research where there is limited funding available. There is also dispute as to whether all medical research scientists, behavioral scientists, and those performing clinical trials of pharmaceuticals should be grouped together and covered by the same general rules of ethics. Where there are questions, designated legal authority should be consulted, since this is an important topic. Various institutions, government agencies, and countries may have different policy, regulatory requirements, and enforcement procedures for research protocols, and this should be known to the research investigator prior to preparing a research protocol.

Some scientists are also subject to advisory committees on ethics oversight that have promulgated additional restrictions on outside consulting, court testimony, and public statements referring to a company's products, services, and community relations efforts.

Caveats

Limits

The rule of law refers to the established limits of human behavior deemed necessary to discourage and prevent harmful social interactions. The limits are defined by (1) common law (case precedents issued by the judiciary), (2) statutes (laws issued by the legislatures), (3) regulations (requirements issued by the government executive agencies), and (4) ethical principles that sometimes

may seem rather remote and poorly enforced. However, there is a social purpose in formulating, promulgating, and making known all of the limits and
requiring compliance by everyone.

Rules

The legal and ethical rules of the road serve to encourage predictable conduct, individual trustworthiness, mutually cooperative behavior, and social
responsibility. There is some inherent ambiguity built into the rules to overcome any tendency for actors to move to the very edge of what is acceptable.
There is no bright line or buffer zone, so resolve questionable choices on the
safe side, well within the limits.

Responsibility

The ethical and legal limits are usually expressed in terms of human behavior, such as intentional acts, negligent conduct, and personal assumption of
the risk. They are often discussed in terms of how others may violate the
rules, but professional responsibility does have a very personal or first-party
focus. There is clear and detailed behavioral guidance available when moral
choices must be made; it is not left to each individual to grossly speculate,
make uninformed decisions, or act in pure self-interest.

Obsession

Human error specialists should be obsessive about things that have the
potential to go wrong, since that attitude helps to prevent mishaps that may
otherwise go unrecognized, be minimized, and are unexpected by others.
The obsession may be fueled by conformance to ethical principles, but moderated by political necessity.

Enforcement

It sometimes may seem that enforcement of ethical rules is lacking and the
rules are stretched to accommodate behavior that is unseemly, unsavory, and
unjust. Enforcement comes and goes in cycles, is often in the form of making
examples of a few violators, and cannot be everywhere every time there is
noncompliance. As in many activities, there is some probability of enforcement when least expected. A more potent remedy is reputation.

chapter thirteen

Discussion questions

The following questions may be used to encourage vigorous classroom discussion of technical human error (at the man–machine interface) and managerial human error (in man–people systems). The form of the questions may add to content knowledge and could provoke many different answers. The pathway to understanding, comprehension, and acceptance of human error and its ramifications is not as easy as it may seem, and group discussions about the questions could be great facilitators. The discussion questions also could be used for content reviews, self-learning, and self-evaluation. Just reading the discussion questions is something of an education; answering them may create a conversion to an advocate for error prevention. If there are appropriate answers, it suggests that the book may be used effectively as a future reference work or checklist reminder. While the examples used in the discussion questions are predominately real-life experiences, the described circumstances have been slightly altered and the names of people, brands, and companies have been omitted for privacy purposes. Some teachers may use the discussion questions for a Socratic method of instruction and leave the rest of the book for reference and checklist purposes.

Smart robots

Manufacturing flexibility is the ability to produce more than one model of a product on an assembly line. It is the ability to be able to rapidly change from one product to another. This usually involves smart robots capable of being programmed for numerous tasks, such as the lifting of heavy parts and locating them for assembly, the application of adhesives or paint, welding, inspection, and the automatic replacement of worn tool heads. *Query:* Will this kind of manufacturing flexibility decrease human errors? Is it a cost reduction for mass-produced items? Could it be a competitive necessity? If the costs of the robots are amortized over a 5-year period and the robots prove reliable, will errors still appear?

Flexible workers

Job flexibility means the ability to move workers from one job to another as needed by the employer. A worker may be assigned one job for 2 hours, then rotated to another job for which he has been trained or be assigned to any other job in the factory. This may require a change in work rules and approval from a labor union if the number of job classifications is reduced for both production workers and the skilled trades. *Query*: Will this form of flexibility serve to decrease or increase error rates? Could it reduce musculoskeletal disorders (repetitive stress injuries)? Is it appropriate for low-production-volume products as a form of skill enhancement to reduce errors?

The unguarded saw

A young student, in her first year at a school of industrial design, carefully observed the instructor as he demonstrated the use of a large table saw. He had removed the metal guard and wood pusher block to improve the visibility of the cutting operation to the students. The pusher block was placed back on the table saw, but it covered the blade only during the initial part of the cutting operation and left the rotating blade exposed, to a half-inch height, during the remaining part of the operation. When the student's time came to cut a piece of wood with the instructor supervising, she moved the pusher block forward with her two hands. As she moved her hands back, they were near, but above the rotating blade. To her surprise, the saw blade cut the tendons in the palm of one hand. *Query*: Did the novice student commit an injury-producing error for which she was responsible during the supervised operation? If a student error, was it intentional, inadvertent, or foreseeable under the circumstances? Was it an instructor error to have removed the guard? Was it a school error for having an unsafe pusher block and a removable guard, and not having a quick-stop brake on the saw blade?

The stairway

At a luxury hotel in Hawaii, several tourists were descending a darkened stairway from the lobby to the darkened dining room. As one person completed the first flight (a series of steps), he turned clockwise in a half landing (the platform between flights), and then started down the second flight to the elevated landing in the restaurant. He misjudged a step location, suddenly lost balance, and began to fall, but his quick movements permitted a recovery of his equilibrium. The error was a near-miss as to what could have been a serious injury. The stairwell was intentionally darkened, the stairs were covered in a dark-color plush carpet, the edges (noses) of the steps were not visually delineated, and the tourist hotel was filled with guests not familiar with the entrance to the restaurant. *Query*: Was this near-miss an error on the part of the guest or an error by the hotel? If the guest was elderly or partially disabled, who would be responsible for a fall down the staircase? Would strip lighting to demark the edge of each stair have provided a visual reference point for the guests descending the stairs? Would a phosphorescent strip on the edge of the step been of help? Did the staircase violate the keep-right principle, used in the U.S., where the staircase ascends to the right (clockwise), imposing less effort on those climbing the steps? Is the keep-right habit important for dimly lit stairs? Since 70% of accidents occur on the top three or bottom three steps, should they be given special attention to prevent error and injury? The last steps included a first view of the interior of the restaurant and the cashier's desk; was this an added visual distraction? What were the errors that could have been a substantial factor in a fall accident at that restaurant?

Acceptable defects

The goal of zero defects is deeply ingrained in corporate cultures, although customers do not expect all products to be free of minor discrepancies. The quality assurance engineer attempts to set reasonable targets for defects that are achievable by their suppliers and by their own manufacturing processes. The target may be expressed as something below a standard given for defects per million opportunities (DPMO). This benchmark is an average of defects for a group of parts or a monthly shipment of parts. For example, a general assembly error rate of 35 DPMO may be assigned to wiring, such as mislocated wires in a cable assembly consisting of 29 wires with 40 termination points. Consolidating the assembly errors with other defects and problems, the acceptable defect rate may be 3000 defects per million cables. Inspection and repairs should reduce the rate before shipment to the final customer. *Query*: Should the error specialist be consulted if there is a defect cluster associated with human performance? Is there a tendency to consider the maximum allowable defects (the DPMO) concept more like an expectancy or a lowered goal, as contrasted with the zero defects goal? Is the consumer aware that many defects in a product are acceptable to a manufacturer, including defects that could induce human error?

Instinct

There are those who attempt to reduce process variations by gut instinct actions. They consider design of experiments, statistical decision making, uncertainty concepts, control charts, the use of probabilities, and error counting to be too difficult, confusing, and costly. They believe everything should be simple and direct. *Query*: Should the gut instinct approach be applied to human error control and reduction? Are the foregoing quality assurance techniques useful enough for human error analyses? What is an acceptable defect rate for the average consumer? Is instinct better than science and engineering in identifying and preventing human error?

Anger

A new employee's medical records included a personality appraisal using the Morey Personality Assessment Screener test. This was a 5-minute self-administered test used in a health care setting. It identified a clinical problem in anger control. *Query*: Should the employee be placed in a job where he could not be provoked into an anger reaction that he could not control? Should this short test be used only to justify a follow-up psychopathology evaluation? What does anger control have to do with human error? Could an easily expressed anger reaction have a disrupting effect on the job performance of other workers?

Fear

The chief executive officer of a corporation needed a very strong negotiator to conduct some important and difficult transactions with another company. He had thought of an executive who had a calm and fearless composure, was cool headed and poker faced, was tranquil regardless of provocation, and was a firm and long-lasting proponent of his issues. However, he was reminded that recent research had shown that those with calm demeanors are willing to take far greater risks than those capable of expressing some fear in choice situations. *Query*: Should the CEO select the poker-faced executive as a negotiator and assume the greater risks involved? Do high-risk choices involve unpredictable outcomes? In what way do risk and unpredictability relate to human error and its control? Do personality tests reveal who is risk aversive and who is a risk taker?

Decision support

In a recent study, more than half of the patients in a hospital suffered adverse drug events. Doctors and nurses seemed to need computer-based decision support services. This would tend to eliminate mistakes from illegible handwriting, perform checks to flag drug allergies and interactions, advise with greater particularity on drugs and dosages, suggest patient monitoring strategies, and provide order sets of drugs and dosages. Many doctors have resisted decision support, indicating that they need no help in remembering what was needed about drugs and treatment. *Query*: Does the high frequency of drug errors indicate a need for decision support, or does a doctor's diagnosis and treatment plan include a great deal of unrecordable subjective experience? Could a hospital culture of safety help to overcome communication errors from doctors who disregard the viewpoints of nurses and ignore machine prompts?

Problem employees

There was a problem employee at a software company who was overly arrogant, uncontrollable, and created ill will among other employees. The supervisor, faced with a troublemaker, decided to recommend the "hot potato" to the manager of another department. There were several subsequent job changes when the other managers avoided confrontation and negative job review or work appraisals. This path of least resistance eventually led to the employee's promotion to a management position. *Query*: Did the job changes serve to reward an incompetent performance? Was this an undesirable personality trait that could have benefited from professional counseling, or was it a more severe personality disorder requiring psychotherapy? In what way could the conduct of this problem employee contribute to human error problems?

Aircraft control

There have been a series of incidents in which commercial aircraft have flown too close to each other, in violation of federal air safety rules. The response of air traffic controllers often has been that they are overworked and understaffed. *Query*: Could such human errors be caused by overwork and understaffing, or is that simply a rationalization (excuse making) that is politically acceptable and could further the general self-interests of air traffic controllers? Could more advanced computer-generated displays and automatic warning devices reduce the errors?

The zoo

There have been many visitor injuries at public zoos that involve animal bites and maulings. *Query*: Does this primarily involve the overtly careless and reckless behavior of the visitors, that is, something approaching an intentional human error? Is such conduct predictable among the wide diversity and intense curiosity of those invited to a public zoo filled with dangerous animals? What is the responsibility of the zoo to prevent errors, by visitors and zoo personnel, that could result in injury?

Pedagogy

Periodically, there have been complaints that traditional pedagogy consisting of lectures, supporting laboratories, and dense factual content may not be the best teaching method. Instructors do use intuition, trial-and-error experience, and personal charisma to enhance learning and student perceptions. There may be a wide gap or disconnect between what is taught, what is learned by memorization, and what results in a general problem-solving capability. There have been suggestions, such as warm-ups (Web-based prelecture material) prior to each lecture, team learning or activity-based instruction, the redesign of classrooms to promote collaborative learning, cross-disciplinary teams to improve communication skills, and the animation of difficult concepts. *Query*: In terms of training or education intended, in part or in whole, to prevent human errors, what type of teaching would be most beneficial? If errors are part of the learning process, do they deserve more research attention? If so, should such research involve the use of brain imaging methods from computed tomography (CT), positron emission tomography (PET), single-photon emission computed tomography (SPECT), or magnetic resonance imaging (MRI)?

Surgical errors

A patient was prepared psychologically in order to instill confidence in the surgeon, then given a small amount of local sedation, and finally the surgeon carefully inserted a catheter (tube) through various tissues to reach and drain the accumulated fluid between the pericardium (membrane surrounding the heart) and the muscular wall of the heart (the myocardium). The surgeon used his past experience as to the catheter insertion depth and feel, but made a judgment error by going too deep and piercing the heart muscle. *Query*: Was this sort of judgment error excusable? Do such errors justify costly computer-assisted surgery where a display monitor visualizes the navigation pathways and tracks the instruments inserted into the body?

Epidural errors

The authors of this book have witnessed the problems in inserting catheters, by distance and feel, in order to administer an epidural anesthetic into the sacral (vertebral) canal without causing injury. The penetration has to be deep enough to go through the dura membrane (sheath covering the spinal canal), but not so far as to cause injury. In an unwitnessed situation, the epidural intervention resulted in aspirating blood. The surgeon then moved further up the spine, but inserted too much of two different anesthesia agents, and the patient's diaphragm was adversely affected. The anesthesiologist in attendance had turned off the alarms because he did not want interference with the ball game to which he was listening. The patient stopped breathing and there was brain damage, life support, and eventual death. *Query*: What life-threatening errors were committed? How could those errors have been avoided? Was the procedure within the standard of care expected by the patient and his kin?

Averaging the errors

Automobile crash testing is necessary to determine the level of safety achieved in a particular design. But humans cannot be experimentally sacrificed for that purpose, so indirect measures have been made through the use of human cadavers and surrogate mannequins (dummies). Thus, some uncertainties remain in areas such as accident reconstruction. The velocities of impacting vehicles may be determined by crush distance and skid marks using commonly accepted computer models (such as CRASH, EDCRASH, SMAC, and IMPACT). For one accident, based on crush distances, the impact speed (delta-V or change in velocity) was 19.6, 20, or 23 mph (based on different accident reconstructionists). The averaging-of-error conclusion, based on all circumstances, was considered to be 20 mph (32.2 km/h) or a barrier equivalent speed of 9 mph (14.5 km/h), with a peak g-loading of 16. Another approach to averaging the errors was the use of two different methodologies pioneered by Raymond M. Brach, a professor of mechanical engineering at the University of Notre Dame. He supplemented the crush distance method with an impulse and momentum method. The momentum is the product of the mass and velocity at the center of gravity. The change in momentum is equal to the impulse applied (the area under a force and time curve). *Query*: Human errors in calculating vehicle velocity are averaged out, but can such errors be eliminated by more precise methodology? Should known calculation and sampling errors (based on extraneous variables and uncertainties) be reported in addition to the normal errors of estimation (the intrinsic variance of the estimate)?

Building evacuation

As a result of the World Trade Center disasters in 1993 and 2001, most people are painfully aware of the problems in escaping (evacuating or egressing) from high-rise buildings. There were 15,400 people evacuated from the Twin Towers on September 11, 2001. There were 2033 others that did not make it (exclusive of emergency responders). Human behavior (egress dynamics) is different in nonemergency, emergency, panic, crowd crush, and deference (permitting lower-floor occupants into the stairwell) situations. Jake Pauls, a well-known architect from Silver Springs, MD, has specialized in building use and safety. He has vigorously recommended building code changes, such as increasing stairwell width from 44 to 56 inches to accommodate the counterflow of ascending emergency responders. He has recommended that handrail continuity be ensured between flights, plus raising the handrails to an ergonomically safe height (above 34 inches) and having better marking where there may be darkness, smoke contamination, and complicated usage. *Query*: Would these code changes reduce the errors that result in unintentional falls in the stairwell? Would photoluminescent egress path marking on the top of the handrails reduce mental confusion and inadvertent crowd crushing? Do traditional oversize and ungraspable handrails induce human error?

Social information

Human decision-making and choice behavior can be strongly influenced by the behavior of others. Social information is obtained by monitoring the cues inadvertently displayed by those having a similar interaction with the environment or cues from those performing more correctly and efficiently. This type of social information is obtained at a much lower cost than trial-and-error learning. The old saying about learning by example has been proven by recent research. In essence, information is extracted from others as a form of behavioral adaptation. *Query*: Does the acquisition of social information serve to reduce errors sufficiently and in a cost-effective manner? Do personality traits, in the giver and the receiver, foster or inhibit this form of learning and adaptation? Is this form of self-learning better for some workers who have personality disorders or regional deficiencies? Will errors be copied?

Pollution

The illegal dumping of waste, on the land and in the ocean, has been a serious environmental and pollution problem. In one vessel pollution case, there was a deliberate discharge of 500 gallons of waste oil into the Columbia River near Kalama, WA. There was an attempt to cover up. Usually, such a discharge is publicly called an accidental error by a member of the ship's crew. In this case, a special bypass pipe was used by the crew members and the conduct was called negligent. Seven other ships, operated by the same company, were found to have bypass equipment that could be used to discharge waste and oil into the ocean from the container vessels. None of the ships had pollution prevention equipment. *Query*: Was this an "error" by an uninformed crew member following a standard practice on the ship? Did the ship owners, who installed the bypass equipment, plan to use it only in acceptable locations elsewhere in the world? Were the fictitious logs, inspected regularly by the U.S. Coast Guard, evidence of intentional conduct by the ship owner rather than inadvertent human error (a violation of company rules and procedures)? Was the bypass of the ship's oil–water separator evidence of a premeditated intentional error?

Sea spray

Many accidents of offshore vessels have involved broken wheelhouse windows that resulted from the impact of sea spray. Traditional hull shapes throw the seawater upwards, slamming into the ship, and then the hull dives deeper. To avoid having the sea creep up the side of the ship onto the bridge deck, there have been hull or bow designs with different flares, slopes, shapes, and widths. Attempts have been made to protect deck equipment from wind, weather, and icing. *Query*: Explain how the improved hydrodynamics, from a better hull design, could effect human error in rough seas and icing conditions? Would the captain of the ship run faster, perhaps too fast, with improved hull design and be potentially chargeable with human error in case of an accident?

Robotized humans

The military services must plan and develop helpful technology in an ever-changing world market for their services. One development has been an exoskeleton or robotized human. Such devices enable foot soldiers to carry heavy loads, move faster, and have extended endurance. For marines, who are carried by ship to combat zones, this enables rapid deployment from ships to land and quick movement inland without waiting for time-consuming logistic support. It also enables quick vertical assault and envelopment of enemy forces or targets well behind the immediate combat zone. There is some reluctance to support any new technology, by top field commanders, because of the fear of technological failures and the uncertainty of anything new and not proven in the field of combat. *Query*: How can the technological uncertainty be reduced, while retaining the element of surprise? How could human errors be minimized, other than by trial and error or special training? Are human errors a prime culprit in the utilization of new and unproven systems?

Energetics and hardening

Some military systems are designed to be self-protective in extreme environments. Naval vessels may be sealed and pressurized to operate in combat areas where there is a presence of radiological, biological, or chemical agents. It may be important for the rescue and treatment of exposed foot soldiers. This may be a vital supplement to individual protection, where garments and respirators can only be used for short-term exposures. The vessels may be hardened to resist energetic devices that may be chemical, light (laser), or microwave in origin. *Query*: Are unusual human errors expected in extreme environments? Should pressurized vessels be available in small numbers to rescue exposed military personnel, or should it become a standard feature for attack warships and large landing craft? Do the discomfort of protective garments, the difficulty breathing with respirators, and the extensive decontamination procedures tend to induce human errors that could be life threatening?

Howitzer misfire

In a very cold climate, a 155-mm howitzer failed to fire. The immediate question was whether it was a dud, a hang fire, or a misfire. The next question was what was the safest procedure in unloading the shell and transporting it to an appropriate location. A great deal of effort has been applied to ordnance safety to ensure safe handling and transportation of explosive shells and to increase fuze reliability, while decreasing cost. *Query*: Would frigid temperatures, in the operational environment, have an effect on recommended procedures and error prevalence? Should special training be provided for those operating large howitzers in very cold weather? In terms of human error, what is the difference between cold and hot, dry and humid, operational environments?

Safety enhancement

There were some problems in shipboard ordnance fuze safety and reliability, particularly with submunitions. The improvement plan was to require dual fuzes and three safety features in each fuze. In addition, there could be size reduction by use of microelectromechanical sensors (MEMS) and safety features. The use of MEMS technology would also reduce assembly and associated error. *Query*: Based on the improvement plan, what were the safety errors? What alternatives were there to the improvement plan?

Aircraft munitions

In the military, there is a need for munitions that can be put in place, subsequently remotely armed, then fired or remotely disarmed, and finally reused. The problems in recovery of previously armed munitions that have been subject to considerable abuse and time delay may subject many people to unacceptable risk. The social responsibility image may resemble the recovery of hundreds of thousands of land mines (nonbiologically degradable) left in place in civilian-occupied areas. *Query*: What human errors could be present in the recovery of munitions that have been fired at a target but failed to explode? How can the remote controls be protected from computer hacking and wireless message intervention? What safety features could be instituted to prevent human error in the arming and firing of the weapon system? What kind of errors are to be expected from civilian populations after warfare terminates?

Robotic systems

Some robotic systems are almost entirely autonomous, be it a welding system or a remotely piloted vehicle used for surveillance, reconnaissance, search-and-rescue operations, live ordnance disposal, or advanced military systems. *Query*: What unique safety problems could arise with a robotic system? What errors could be introduced into the system function? What errors might affect downtime, including scheduled maintenance and necessary repair procedures?

Reactive chemicals

It has been well publicized, as a result of the 1984 Bhopal toxic gas release, that there has been an emphasis on inherently safe design in chemical plants. This means that lethal reactive chemicals that are intermediaries are made and used continuously within the manufacturing processes. The on-site storage (stockpiles) of such chemicals is thus eliminated or sharply reduced. *Query*: Does this reduce the possible errors in the operation of safety valves and chemical transfer hoses and lines? Does it help make managers more vigilant and aware of uncontrolled chemical reactions? Is there still a need for independent process safety oversight? Is the best hierarchy one that starts with hazard removal, then passive process controls, active controls, and, last, operating procedure safeguards?

Decisions

There has been considerable discussion on how persons and organizations make decisions. One assumption is that man is a rational and reasonable person who can quickly make simple decisions or even complex decisions based on what limited information is then available. It may seem to be decisions without too much thought. Another assumption is that humans recognize all the choices, assess their consequences, and evaluate the weight of evidence in favor of each before making a decision. This seems time-consuming and complex. The third assumption is a "good enough" or reduced content approach that considers a limited amount of relevant information. *Query*: Since a common complaint among first-line workers is that they do not have enough information, does this mean that more communicative work supervisors might reduce error? What does this mean in terms of job enrichment? What is the balance between good-enough decisions and better-informed decisions in terms of human error?

The tour route

Visitors from outside or within a company are often given a short or long tour, with escorts or guides that expound on rehearsed favorable platitudes and eschew known problems. Traveling the tour route may create a favorable but incorrect impression or image. The less traveled pathway may reveal contentious workplaces, violations of company rules, and sloppy work practices. *Query*: Given the common attempts to cover error-producing situations, what could be done to discover human error and correct it? Could an unscheduled, unannounced, and unexpected visit be conducted in a civil, friendly, and mutually beneficial basis? Are defensive reactions symptomatic of known error sources?

Hole in the ground

A slightly overweight middle-age man was jogging in his residential neighborhood during evening hours. It was dark, with no street lights, so he ran on the concrete sidewalks. Suddenly he tripped and fell, suffering painful injuries. He looked back and saw a red plastic warning cone on the ground near the sidewalk, suggesting someone knew of a danger. There had been construction at a residence and the construction vehicles had run over the concrete sidewalk, crumbling it, depressing the adjacent ground, and causing an upheaval of two water meters. An agent of a private building inspector had seen the water meters out of position and had notified the building department at city hall. They placed the red plastic cone over a water meter in the unlit ground area. They did not attempt to fill the depressed ground area in the property set back. They did not attempt to use an asphalt mixture that could be applied cold and tamped down to temporarily repair the edge of the concrete sidewalk. *Query*: What was reasonably necessary to control the following human errors: construction truck drivers running over the sidewalk and the water meters creating damage, municipal employees placing a red cone in an unlit ground area, the failure to fill in the hole and make a temporary repair of the sidewalk, failure to restrict the use of the sidewalk, and a jogger running in an area so dark that he could not see the unreflectorized red cone adjacent to the sidewalk. Should the city, contractor, or inspector have had a safety plan in effect to prevent such injuries, or was this a rare and unpredictable situation? Could this be a simple situation where a so-called fat person was running in a careless manner with a seemingly litigious attitude, a contractor needing access and using commonly used methods, an underfunded municipality having limited services, and a homeowner without fault? What are the commonsense rights and responsibilities of each of the participants in such a situation?

Close call

The driver of automobile A was traveling at a moderate speed in the number 1 (fast) lane of a broad urban roadway. As the driver approached an intersection, he prepared for a lane change into a left-turn pocket of the traffic signal-controlled intersection. The driver turned on the left turn signals (lights) and glanced in the rearview mirror mounted on the door to see if there was a clear pathway. Seeing nothing, he started to turn. Meanwhile, the driver of automobile B, who was following automobile A, decided to prematurely move to the left and accelerate to quickly pass the traffic ahead of him. The result was a near-miss of a two-vehicle sideswipe, a miss avoided only by a fast recovery (a quick return to the original lane) by the driver of automobile A. In this case, there would be no record of a near-miss, close call, or propensity for collision or accident. *Query*: Was this a learning experience about human error? What human errors were committed, why, and how could they be prevented in the future? Was there some commonsense fault? If a traffic officer witnessed the near-miss, could he have attempted enforcement of the traffic rules? Do such human errors occur frequently? Do they result in a significant number of injury-producing collisions? Will they continue unabated? What is the best remedy for driver error?

Regional differences

Large manufacturing and assembly plants are located where labor costs are low, local governments offer considerable subsidies, suppliers are nearby or are willing to relocate to an adjacent industrial park, the needed skill levels can be found or trained in the local population, and appropriate levels of production quality can be reasonably expected. Some manufacturers claim that there are significant regional differences in workforce literacy, aptitudes, and motivation. *Query*: If such beliefs about regional differences in the workforce are true, how would this affect possible manifestations of human error?

Could proper extended training minimize any such differences? Could this be an incompatibility between company cultures and local cultures that could be resolved? Do regional differences affect the kind of error counter-measures that could be effectively implemented?

The 5% rule

In the design of large complex systems, approximately 5% of the budget has been allocated to human system integration. This includes human factors, system safety, and design safety. Typically, this funding has been spent on other subject areas, including training requirements. Often this grouping of budget funds has been an attempt to merge disciplines and provide management flexibility. *Query*: Are there advantages or limitations to treating such activities as a cohesive whole, rather than independent domains with clear assignments of responsibility and authority? Is 5% sufficient or just an arbitrary rule based on insufficient knowledge of the methodologies, perspectives, and culture of each discipline? Is "blended" a better process for combining theory and practice? Should a budget be allocated by a project engineer in conformance with performance and projected needs? Should error prevention be a line or a staff function?

One keystroke from danger

In software systems development, there is a system safety rule that only one keystroke or movement should eliminate a possible harmful result. Software developers for military systems have publicly published manuals describing many design safety techniques compatible with reliability and system safety analytic techniques. *Query*: What is meant by one keystroke from danger? Is this particularly relevant to other human errors in the operation of a system? Is it sufficient? Could this be related to dead core codes? Is a "quick stop" a way to prevent errors?

Things happen

A serious safety problem arose on a piece of complex equipment used in a system context. In an attempt to get the problem resolved, contact was made with the field representative of the manufacturer of the equipment. The immediate response was "things happen," "someone must have done something wrong," and "we don't have enough accurate facts to assess the problem." *Query*: What should be done in view of the field representative's statements? If the problem became manifest after the expiration of warranties, what should be done? If it were a critical safety problem, what should be done? Should the field representative institute his own investigation of the human error? Is it true that the greater the evasive rationalization, the greater the effort needed to isolate the real error problem?

Stress and mental confusion

A design engineer attended a conference in a remote city. He had been supplied with a street map marked with certain building locations (reference points). He did not know that some of the reference points were incorrectly located and marked by his office staff. Arriving late, he decided to walk to a good restaurant. Coming back to his hotel he became lost. He wandered for more than 2 hours, in a hot sweltering climate. The navigation errors, engendered by the false reference points on the map, created mental confusion and resulted in poor decision making. His choice behavior became erratic and essentially nonrational or emotional. The increasing physical and mental stress increased the error rate. *Query*: In what manner and why is choice behavior adversely affected by high levels of stress? Is this a physiological or mental problem? Since machine operators may encounter high stress, particularly in emergency situations or events of high criticality, what should be done to prevent error and ensure error-free performance? What are the short-term and long-term effects of contingency training on foreseeable human error? Is it true that the application of torture also involves creation of high stress, loss of reference points, inducing confusion, and the creation of opportunities for error?

Space station

The design of the International Space Station had a life support system that included carbon dioxide (CO_2), oxygen (O_2), and water (H_2O). The level of consumption of each would vary with the number of crew members. The best available probability data were used to formulate a mean time between failure (MTBF) and a mean time to repair (MTTR) for the life support equipment. As such core systems are reduced in weight, the payload can be increased. The overall design attempted to eliminate numerous hazards, ranging from crew members' eye irritants to an unwanted navigational system shutdown. *Query*: Since the astronauts received intensive training, had extensive experiences with simulated flight conditions, and were well aware that crew teamwork was a vital defense against disaster, would significant human error be expected during flight operations? During assembly, test, and movement of the space shuttle on the ground, would there be an equal expectation of human error? What major safeguards, if any, were eliminated for the space station project? Would emergency rescue of astronauts, if possible, include unanticipated human error? As the space station moves toward obsolescence, do errors increase with new flight crews?

Medical mistakes

Television news anchors, in 2005, headlined a "million medical mistakes" a year in the U.S. About the same time, a Canadian medical researcher indicated that Canada had a similar problem, but made recommendations that included standardization of procedures so that every physician does things in the same way and in an accepted manner. He advocated a systems approach to patient care, building a team spirit with equivalence of ranks and functions, the creation of a learning culture, and better feedback, which would improve situation awareness of medical conditions. *Query*: Would these general policy changes have a beneficial effect in preventing human error (medical mistakes)? Since most physicians prefer autonomy as a form of professionalism and this is a valued and embedded belief, should this individualism be changed to standardize medical services? Would process mapping help in selecting processes that need reengineering to forestall medical errors? Is there a parallel between the professional practices of medicine and the professional practice of human error engineering?

Oil refinery shortcomings

Prior to the start-up of an octane-boosting processing unit, there was an awareness that repairs were needed and that some key alarms were not working. A work request was approved by a company manager indicating that the repairs would be deferred until after the start-up. An explosion occurred during start-up that killed 15 people and injured another 170. Earlier start-ups had often exhibited abnormally high pressure and temperature readings in that processing unit. *Query*: Was there a managerial error in approving the start-up operation? Was there a sufficient warning by virtue of high pressure and abnormal temperature readings during prior start-ups that suggested that something was wrong? Were the alarm problems indicative of some form of human error? Was there a safety culture problem that induced human error?

Unknown probability of catastrophic events

There have been many disasters where the probability of occurrence was very low or unknown. Examples include major earthquakes, devastating tsunamis, public health epidemics, asteroid impacts, terrorist attacks, unexpected high-energy explosions, unusual extensive fires, unprecedented floods, and costly recalls. The occurrence of a catastrophic event being of an unknown probability, or having never occurred before and seeming highly unlikely, has often justified ignoring the problem and discounting the need for precautionary measures. *Query*: Would it be a human error to decide not to tell those who might be exposed to a tsunami that a sudden recession in the oceanfront water might signal danger? Should such information be conveyed by emergency broadcasts, sirens, telephone calls, or a planned emergency response system? Is relocation feasible? If no action is taken, is that a political human error, a governmental error, or an error on the part of those knowingly exposed to great danger? Should cost justification be based on possible deaths, the cost of precautions, and the possible reduced death toll? Is a once-in-100 years disaster something so distant in time that it reasonably could be ignored, or is that a judgment error?

Attention

The selective attention of a person depends on his expectations and goals. The person may be subjected to an overwhelming flow of information, but responds to only that which is behaviorally relevant to those expectations and goals. There may be a selective activation of the lateral intraparietal area, modulated by the working memory and executive control in the prefrontal cortex. That brain activity is important where human attention is needed. Selective attention (alert focus) is important in many human activities. *Query*: What may bias or voluntarily shift attention to or from the desired salient information in the visual field? If a work task needing attentional control is subject to distraction, would the distraction itself or the failure to detect be an error? When the locus of attention (the target) may not be remembered, how can error be prevented? Is human error due to inattentiveness the fault of the brain, the person, the situation, or the task?

Imitation

An important mechanism of learning occupational and social skills is by observation of others. As the imitator watches the actor there is specific neural activity, the basis for which may involve different mechanisms, depending upon the observed behavior and any spatial or symbolic cues involved in the process. It could be the left inferior frontal cortex for the general performance of a task and the right superior parietal lobule for the exact movements. Broca's area plays a significant role in language acquisition in learning by imitation. The identification of specific brain areas has been made possible by the use of recent techniques such as functional magnetic resonance imaging (fMRI) in the living body (*in vivo*). *Query*: Are such imaging techniques more objective and the results more real, as compared with prior learning theory and behaviorism? Do such brain mapping and cortical mechanisms have immediate or future promise for human error prevention? Where there is learning by imitation, how could human errors be repeated by the imitator or incorrectly visualized and integrated into a work routine?

Stress-induced alcoholism

Stressful life events may increase alcohol (ethanol) drinking among many people. Above a certain blood alcohol level, driving while drinking results in such an abundance of human errors that driving is prohibited by almost all vehicle codes. Most people do not increase their alcohol intake under repeated stressors, such as social defeat and physical struggling. Other people, who may have problems in genetic predisposition and hormone release mechanisms affecting behavioral response to stress, have a markedly different outcome. Their alcohol drinking may be an attempt to cope with stress after the failure of maladaptive behavior, but repeated stress progressively increases alcohol consumption, a condition that generally persists throughout their lifetime. *Query*: For purposes of human error control, what job assignments should or should not be given to chronic alcoholics, either in remission or during relapse? What are the effects of employment laws, discrimination laws, and labor laws, in your locality, in reference to recovering alcoholics? How does the error-producing behavior of an alcoholic compare with other addictive disorders described elsewhere in this book? Would your answers be different if there were new drugs to supplement traditional psychotherapy, abstinence support groups, and the use of Antabuse (which produces nausea if drinking occurs)? Such new drugs may include those that blunt cravings, help maintain abstinence, and lessen relapses. These new drugs may or may not have side effects that influence memory, concentration, and grogginess. If 8% of the adult population in the U.S. suffer alcoholic dependence or abuse, would that constitute an industrial enterprise problem in terms of human error or a public health problem that should be of government concern?

Cultural norms

The human error specialist should be concerned with subgroup cultures, which include group practices and beliefs passed on from one generation to another. Culture, in general, is a pattern of socially learned traditions and behaviors. The individual is taught a predictable behavior repertoire acceptable by his social group. This is analogous to the human conduct guidelines provided by statutory and common law, which are a necessity for predictability in complex social settings and to ensure harmonious personal interactions. A transgression of the human conduct limits or boundaries of the law may result in civil and criminal penalties. *Query*: How are subgroup norms enforced? Should violations of cultural norms or of the common law be considered human errors, social errors, legal errors, or cultural errors? Does socially transmitted behavior ensure a collective norm that reduces variant behavior but increases error in a broad community sense? Will the Internet reduce errors associated with cultural norms that differ because of geographical distance, educational content, past local experience, and new expectations?

Cues and memory retrieval

Humans have the capacity to retrieve complete memories on the basis of an incomplete set of cues or degraded fuzzy inputs. This pattern completion function is essential for learning and accurate memory recall. It is a function of the associative memory network in which memories are stored by synaptic changes in the neural circuitry of the hippocampus. For work performance without errors, there must be accurate memory storage, retrieval, and pattern completion. *Query*: Are human errors expected in a full-cue situation? In a healthy biological system, would partial-cue presentation result in complete associate memory recall? Where there has been reactivation or reinforcement of the pattern, will there be less error in responses to memory recall? If there is a more robust input to storage, would there be less cellular inhibition of the possible memory trace?

Cognitive memory

The cognitive memory system is a brain-wide distributed network that includes the frontal cortex, temporal cortex, medial temporal lobe, and a sub-system in the parietal cortex. Memory may be fact (semantic-like general knowledge), based in the temporal cortex region, or event (episodic-like past knowledge), based in the medial temporal lobe. Together they form the declarative or explicit memory (for facts and events). The frontal cortex manipulates and organizes the information so that it can be remembered. The implicit memory modifies behavior without awareness. *Query*: Are such explanations of cognitive memory too confusing and too tentative to be of practical use by the human error specialist? Would it be more valuable if it were possible to have an intervention approach, such as a drug injection or reversible gene insertion, that improves memory retention? What are the current effects of illegal drugs on cognitive errors, as described in the text of this book?

Linear programming

There are various flow optimization mathematical models or methods including operations research to maximize flow (such as traffic) and linear programming to maximize or minimize a linear function (a throughput). There may be an attempt to minimize risk by partitioning the problem into resolvable subproblems to find optimum solutions. In such analyses, the human component of the system may be treated as another independent variable to be included in an overall system analysis. *Query*: Does a human component in a system act as a linear or nonlinear variable? Are machine errors, such as those due to wear or material inclusions, more or less predictable than humans who may introduce unanticipated error into the system? If there is no mathematical theory that can find a global optimum design for nonlinear functions (such as design variables and human variables), can approximations of the linear sequence be utilized? Are industrial engineers, who are familiar with linear programming, more able to include human error in their system optimization efforts?

Safe exposure levels

A threshold limit value (TLV) is considered an exposure level to toxic agents at which most people suffer no harmful effects when working 8 hours a day continually. Some of these "safe" exposure limits are periodically changed upward or downward depending on more recent research findings. Recently, a National Research Council (NRC-USA) report indicated that the risks of low-dose radiation are small, but there is no safe level. The extrapolation of radiation damage to cellular DNA utilized a linear no-threshold (LNT) model. This was confirmed in studies of nuclear workers and people exposed to medical radiation. There have been challenges to the LNT model that state that a little radiation is harmless and may actually stimulate DNA repair processes (known as hormesis). *Query*: Did those professionals and technicians who relied on the accuracy of TLVs that later changed commit a judgment error? If they did not incorporate a safety factor (similar to those utilized by mechanical engineers for uncertainties, unknowns, and possible changes), did they commit an error? If someone was harmed by a toxic exposure, well within an acceptable TLV, did someone commit an error? Do such error concepts apply to terms similar to TLV, such as PEL (permissible exposure level), MPC (maximum permissible concentration), and MPD (maximum permissible dose)? If errors, were they acceptable errors based on the then available knowledge, the state of the art, and the peer acceptance of such decisions?

Multiple causation

A human error mode was found in a manufacturing process. An errorectomy was performed. However thoughtfully and carefully done, the error reappeared. Apparently, there was more than one cause for the error mode. Fortunately, the follow-up of the first errorectomy revealed that the error problem remained unabated. Multiple causation may suggest that the original root cause was not detected and corrected. There are professionals who insist that every event has multiple causes even if supervising authorities are interested only in *the* cause. Causality itself is subject to change as illustrated by Freudian searches for childhood trauma or fantasy, then strict behaviorism with a stimulus–response approach, and recently the emphasis on neuroscience. This book emphasizes the practical prevention of error rather than the earlier emphasis on theory and models. Time and experience change concepts of causation and remedy. *Query*: Could the first remedy have been ineffective because it did not counteract the actual error mode? Does this scenario suggest the real need for follow-up and subsequent audits? Regardless of the theories utilized, if the practical or desired objective of preventing or correcting the source of human error is achieved in a technically and cost-effective manner, does theory matter? Is the concept of excising error better than countering it?

Attitude matters

There are those who believe that the primary cause of automobile accidents is driver error. They strongly state their conviction that vehicle drivers are inept, lack driving skills, can be careless, are often inattentive to the driving task, and are frequently intoxicated. Some people fault the vehicle designers and engineers with claims that there have been tens of thousands of preventable deaths and injuries from defects and design deficiencies. Other people blame the roadway system, the vehicle manufacturers, the police for failure to ensure compliance with the rules of the road, or the federal government for not mandating available safety devices. Such beliefs and opinions can escalate into strong convictions, personal attitudes, biases, and prejudices. *Query*: If an automobile accident is investigated and reconstructed, would an investigator's deeply embedded personal attitudes regarding who is generally at fault have an effect on how the facts are perceived, interpreted, and categorized? Could a design engineer and his supervisor, faced with charges of possible product deficiencies, develop a defensive posture and then a negative attitude toward suggested design improvements? Assume that a corporate manager establishes a technical committee on competitiveness, a council on long-range product objectives, and a think tank on human error and productivity. Should negative attitudes be an exclusionary factor in staffing such management advisory groups? If the company has been plagued by strategic missteps and resultant tactical errors, is some form of a quasi-independent positively oriented attitude desirable? Could a human error search-and-destroy effort be successful if conducted by naysayers, those with negative attitudes, those with dismissive passive loyalties, or those who reject the Socratic method of interrogation?

References and recommended reading

AASHTO, *A Policy on Design of Urban Highways and Arterial Streets*, American Association of State Highway and Transportation Officials, Washington, DC, 1973.

Abboud, L., Mental Illness Said to Affect One-Quarter of Americans, *The Wall Street Journal*, June 7, 2005, pp. D-1 and D-7.

AC65-9A, *Airframe and Powerplant Mechanics General Handbook*, Revision, Federal Aviation Administration, U.S. Government Printing Office, Washington, DC, 1976.

Adams, M.J., Tenny, Y.J., and Pew, R.W., Situation awareness and the cognitive management of complex systems, *Human Factors*, 37, 85–104, 1995.

AF Manual 51-37, *Instrument Flying*, Department of the Air Force, Washington, DC, 1971.

Alexander, G.J. and Lunenfeld, H., 1999, Positive guidance, in *Warnings, Instructions, and Technical Communications*, Peters, G.A. and Peters, B.J., Eds., Lawyers & Judges Publishing Co., Tucson, AZ, 1999, chap. 10.

Allard, K., *Command, Control, and the Common Defense*, revised edition, National Defense University, Washington, DC, 1996.

ANSI B11.TR3-2000, *Risk Assessment to Estimate, Evaluate and Reduce Risks Associated with Machine Tools*, American National Standards Institute, New York, NY, 2000.

ANSI Z535.4, *Product Safety Signs and Labels*, American National Standard, National Electrical Manufacturers Association, Rosslyn, VA, 1998.

ANSI/ISO/ASQ Q9004-2000, *Quality Management Systems: Guidelines for Performance Improvement*, ASQ Quality Press, Milwaukee, WI, 2002.

ASME Safety Division, *An Instructional Aid for Occupational Safety and Health in Mechanical Engineering Design*, American Society of Mechanical Engineers, New York, 1984.

ASTM F 1166-88, *Standard Practice for Human Engineering Design for Marine Systems, Equipment and Facilities*, American Society for Testing and Materials, Philadelphia, 1988.

ASTM 1337-91, *Standard Practice for Human Engineering Program Requirements for Ships and Marine Systems, Equipment, and Facilities*, American Society for Testing and Materials, Philadelphia, 1991.

Baerwald, J.E., Ed., *Transportation and Traffic Engineering Handbook*, Institute of Traffic Engineers, Prentice Hall, Englewood Cliffs, NJ, 1976.

Belmont Report, *National Commission for the Protection of Human Subjects of Biomedical and Behavioral Research. The Belmont Report: Ethical Principles and Guidelines for the Protection of Human Subjects of Research,* OPRR Reports, U.S. Government Printing Office, Washington, DC, 1979.

Berkow, R., Beers, M.H., and Fletcher, A.J., *The Merck Manual of Medical Information,* Merck Research Laboratories, Whitehouse Station, NJ, 1997.

Bruner, R.F., *Deals from Hell,* Wiley, New York, 2005.

BS 2771, *Electrical Equipment of Industrial Machines,* British Standards Institution, London, 1986.

BS 5304, *Safeguarding of Machinery,* British Standards Institution, London, 1975.

BS 5304, *British Standard Code of Practice, Safety of Machinery,* British Standards Institution, London, 1988.

BS 8800, *Occupational Health and Safety Management Systems Guidance Standard,* British Standards Institution, London, 2004. (Provides guidance on risk assessment and control that is acceptable to the Health and Safety Executive and Health and Safety Commission, U.K.)

BS EN 292, *Safety of Machinery: Basic Concepts, General Principles for Design,* 1992.

BS EN 60204, *Safety of Machinery: Electrical Equipment for Industrial Machines,* British Standards Institution, London, 1985, 1993.

BS EN ISO 9001, *Quality Management Systems: Requirements,* British Standards Institution, London, 2000.

BS EN ISO 14001, *Environmental Management* Systems, British Standards Institution, London, 2004.

BS EN ISO 19011, *Guidelines for Quality and/or Environmental Management Systems Auditing,* British Standards Institution, London, 2002.

BS-OHSAS 18001, *Occupational Health and Safety Management Systems: Specification,* British Standards Institution, London, 1999, amended 2002. (The audited standard for management systems, somewhat similar in function to ISO 9001 and ISO 14001, and compatible with ISO 9001 (2000, quality management systems) and ISO 14001 (1996, environmental management systems).)

BSI-OHSAS 18002, *Occupational Health and Management Systems: Guidelines for the Implementation of OHSAS 18001,* 2000, amended in 2002. (The guidance standard.) British Standards Institution, London.

Burton, T. and Pulls, M.G., 5 Defibrillator Models, *The Wall Street Journal,* June 27, 2005, p. B-2.

Carmichael, C., *Kent's Mechanical Engineers' Handbook, Design and Production Volume,* 12th ed., Wiley, New York, 1950, pp. 8-05 to 8-06.

Chapanis, A., Garner, W.R., and Morgan, C.T., *Applied Experimental Psychology, Human Factors in Engineering Design,* Wiley, New York, 1949.

Chemical Manufacturers Association, *Risk Analysis in the Chemical Industry,* Government Institute, Rockville, MD, 1985.

CIOMS (Council for International Organizations of Medical Sciences), *International Guidelines for Ethical Review of Epidemiological Studies,* CIOMS, Geneva, 1991.

CIOMS (Council for International Organizations of Medical Sciences), *International Ethical Guidelines for Biomedical Research Involving Human Subjects,* CIOMS, Geneva, 1993.

COM, *General Product Safety;* see Commission Amended Proposal for a Council Directive Concerning General Product Safety, COM (90) 2.59 final-SYN 192, 90/C 156/07, submitted by the Commission on June 11, 1990, *Official Journal of the European Communities,* No. C156/8014, May 27, 1990.

Council of Europe, Recommendation R(90)3 concerning medical research on human beings, The Council of Europe, Strasbourg, 1990.

Council of Europe, *Convention for the Protection of Human Rights and Dignity of the Human Being with Regard to the Application of Biology and Medicine: Convention on Human Rights and Biomedicine*, The Council of Europe, Strasbourg, 1996.

Cullen, W.D., 1990, *The Public Inquiry into the Piper Alpha Disaster*, Her Majesty's Stationery Office, London, 1990.

Davies, P., First State Hospital Report Card Issued, *The Wall Street Journal*, January 20, 2005, p. D5.

Declaration of Helsinki, *British Medical Journal*, 313, 1996, pp. 1448–1449.

Deroche-Gamonet, V., Belin, D., Piazza, P.V., Evidence for addiction-like behavior in the rat, *Science*, 305, 1014–1017, 2004.

Dewey, S., Start with a good foundation, *Crane Works*, 14, 16, 2005.

Diagnostic and Statistical Manual of Mental Disorders, 4th ed., Text Review (DSM-IV-TR), American Psychiatric Association, Arlington, VA, 2000, Section 292.9, pp. 241–250.

Dooren, J.C., One once-monthly drug may ease challenge in treating alcoholism, *The Wall Street Journal*, April 8, 2005, p. D-4.

ECPD, Engineers shall hold paramount the safety, health, and welfare of the public in the performance of their professional duties, Section 1, *Suggested Guidelines for Use with the Fundamental Canons of Ethics*, Engineers' Council for Professional Development, New York; also in the *Fundamental Canons, Code of Ethics of Engineers*, ECRD, New York, 1974.

EEC, Council Directive of June 14, 1989, on the Approximation of Laws of the Member States Relating to Machinery, 89/392/EEC, *Official Journal of the European Communities*, No. L183/9-14, June 29, 1989. (Also see Annexes I to VII, *Official Journal of the European Communities*, No. L183/15.32, June 29, 1989.)

Eisenbergër, N.I., Lieberman, M.D., and Williams, K.D., Does rejection hurt? An fMRI study of social exclusion, *Science*, 302, 290–292, 2003.

Eisner, L., Brown, R.M., and Modi, D., Risk management implications of IEC 60601-1, 3rd ed., in *Compliance Engineering, 2005 Annual Reference Guide*, Canon Communication, Los Angeles, CA, 2005, pp. 116–121.

EN 574, *Two-Hand Control Devices — Functional Aspects — Principles for Design*. British Standards Institution, London, 1997.

EN 954, *Safety of Machinery: Principles for the Design of Fail-Safe Control Systems*, British Standards Institution, London, 1997.

EN 1050, Safety of machinery: risk assessment, *Principles of Risk Assessment*, European Norm, EU, British Standards Institution, London, 1997.

EN 50100, *Safety of Machinery: Electrosensitive Safety Devices*, European Union, 1996.

Endsley, M.R., Measurement of situation awareness in dynamic systems, *Human Factors*, 37, 65–84, 1995.

ENTC 114/N 258, *Switching Devices*, European Committee for Electrotechnical Standardization, CENELEC, Brussels, Belgium.

Ericson, C., *Hazard Analysis Techniques for System Safety*, Wiley, New York, 2005.

Ericson, C.A., Risk matrix values and the hazard risk totem, *Journal of System Safety*, 40, 6, 2004.

Firenze, R.J., Establishing the Level of Danger, *Professional Safety*, 50, 1, 28–35, 2005.

Forlin, G. and Appleby, M., *Corporate Liability*, Butterworth Tolley, Lexis Nexis, London, 2004.

Foster, K.R., Vecchia, P., and Repacholi, M.H., Science and the precautionary principle, *Science*, 288, 979–981, 2000.

Fruchter, B., *An Introduction to Factor Analysis*, Van Nostrand, New York, 1954.

Garret, J.W. and Kennedy, K.W., *A Collation of Anthropometry*, 2 vols., Aerospace Medical Research Laboratory, Wright-Patterson Air Force Base, OH, 1971.

Goerth, C.R. and Peters, G.A., Product hazard communications, in *Packaging Forensics*, Stern, W., Ed., Lawyers & Judges Publishing Co., Tucson, AZ, 2000, chap. 6, pp. 43–50.

Hargreaves, E.L., Rao, G., Lee, I., and Knierim, J.J., Major dissociation between medial and lateral entorhinal input to dorsal hippocampus, *Science*, 208, 1792–1794, 2005.

Hasenfratz, M. and Battig, K., No psychophysiological interactions between caffeine and stress?, *Psychopharmacology (Berlin)*, 109, 283–990, 1992.

Health and Safety at Work Act (U.K.), Elizabeth II, Her Majesty's Stationary Office, London, 1974, chap. 37, available at http://www.healthandsafety.co.uk/haswa.htm.

Hendrick, H.W. and Kleiner, B.M., *Macroergonomics: An Introduction to Work System Design*, Human Factors and Ergonomics Society, Santa Monica, CA, 2001.

HFES, Code of Ethics, in *Human Factors and Ergonomics Society Directory and Yearbook, 2004–2005*, HFES, Santa Monica, CA, 2004, pp. 358–360.

Hileman, B., Ethics of Human Dosing Trials, *Chemical and Engineering News* August 1, 2005, pp. 27–29.

Hirstein, W., *Brain Fiction: Self-Deception and the Riddle of Confabulation*, MIT Press, Cambridge, MA, 2005.

Hubbard, R.B. and Minor, G.C., Eds., *The Risks of Nuclear Power Reactors*, Union of Concerned Scientists, Cambridge, MA, 1977. (Provides an example of a critical review of validity of a risk assessment, in this case the *NRC Reactor Safety Guide*, WASH-1400 (NUREG-75/014), in a public attempt at risk reduction.)

Huber, D., Veinante, P., and Stoop, R., Vasopressin and oxytocin excite distinct neuronal populations in the central amygdala, *Science*, 308, 245–248, 2005.

IEC 60601-1, *Medical Electrical Equipment. Part 1. General Requirements for Basic Safety and Essential Performance*, 3rd ed., International Electrotechnical Commission, Geneva, approved August 2004, published April 2005.

ILO-OSH, *Guidelines on Occupational Safety and Health Management Systems*, International Labor Organization, Geneva, 2001.

International Classification of Diseases, 9th ed., *Clinical Modification (ICD-9-CM)*, 6th ed., National Center for Health Statistics, CDC, Hyattsville, MD, 2004.

International Conference on Harmonization of Technical Requirements for the Registration of Pharmaceuticals for Human Use, *Guidance Document*, Health Products and Food Branch, Minister of Health, Health Canada, Ottawa, Ontario, 1997. (Recommended for adoption by the European Union, Japan, and the U.S.)

International Statistical Classification of Diseases and Related Health Problems, 10th revision, 2003 version, World Health Organization, Geneva, 2003.

Ioannidis, J.P.A., Schmid, C.H., and Lau, J., in *Sourcebook on Asbestos Diseases: Right to Care, Epidemiology, Virology, Gene Therapy and Legal Defense*, Vol. 18, Peters, G.A. and Peters, B.J., Eds., Lexis, a division of Reed Elsevier, Charlottesville, VA, 1998, chap. 4.

ISO 14971, Application of Risk Management to Medical Devices, International Organization for Standardization, Geneva, Switzerland, 2000.

ISO 9001, *Quality Management Systems: Requirements*, 1994 and 2000; also see BS EN ISO 9001 (2000) with same title.

James, J.E. and Gregg, M.E., Hemodynamic effects of dietary caffeine, sleep restriction, and laboratory stress, *Psychophysiology*, 41, 917–923, 2004.

Karwowski, W. and Marras, W.S., *Occupational Ergonomics Handbook: Fundamentals and Assessment Tools for Occupational Ergonomics*, 2nd ed., 2 vols., CRC Press, Boca Raton, FL, 2005.

Kilman, S., Alert in Latent Mad-Cow Case Was Delayed by a Misdiagnosis, *The Wall Street Journal*, June 27, 2005, p. A-2.

Leitnaker, M.G. and Cooper, A., Using statistical thinking and designed experiments to understand process operation, *Quality Engineering*, 17, 279–289, 2005.

Levine, R.J., *Ethics and Regulation of Clinical Research*, Yale University Press, New Haven, CT, 1986.

Lowrance, W.W., *Of Acceptable Risk*, William Kauffman, Los Altos, CA, 1976. (Provides a general review of risk concepts and safety decision making.)

Management of Health and Safety at Work Regulations (U.K.), 1992.

Mannan, M.S., The Mary Kay O'Connor Process Safety Center: The First Ten Years, *Hydrocarbon Processing*, Gulf Publishing, Houston, TX, 2005, pp. 65–71.

Manual on Uniform Traffic Control Devices, Federal Highway Administration, U.S. Government Printing Office, Washington, DC, 1988. (Also ANSI D6.1e-1989.)

Maslow, A.H., A theory of human motivation, *Psychological Review*, 50, 370–396, 1943.

Maynard, H.B., Ed., *Industrial Engineering Handbook*, 3rd ed., McGraw-Hill, New York, 1971.

McCormick, E.J., *Human Engineering*, McGraw-Hill, New York, 1957.

McGregor, D., *The Human Side of Enterprise*, McGraw-Hill, New York, 1960.

Meckler, L. and Pasztor, A., Regulators Reveal Near-Collisions of Planes in Dallas, Los Angeles, *The Wall Street Journal*, June 24, 2005, p. B-2.

MIL-S-38130, *Safety Engineering of Systems and Associated Systems, and Equipment, General Requirements For*, U.S. Air Force, September 30, 1963. (Preceded MIL-STD-882D, Standard Practice for System Safety, 2000; also see *Inherently Safer Chemical Process: A Life Cycle Approach*, Center for Chemical Process Safety, AIChE, New York, 1996.)

MIL-STD-882D, *Standard Practice for System Safety*, U.S. Dept. of Defense, Washington, DC, February 10, 2000.

Murff, H.J., Gosbee, J.W., and Bates, D.W., Human factors and medical devices. In Evidence Report. Technology Assessment No. 43, Making Health Care Safer: A Critical Analysis of Patient Safety Practices, U.S. Dept. Health and Human Services, Rockville, MD, 2001, 2004, available at http://www.ahrq.gov/clinic/ptsafety/chap41a.htm.

Navakatikian, A.O. and Grigorus, A.G., Effect of aminazine, caffeine, and the intensity of mental stress on psychophysiological functions and work effectiveness of humans, *Fiziologicheskii Zhurnal*, 35, 6, 79–83, 1989 (in Russian).

Nie, Z., Schweitzer, P., Roberts, A.J., Madamba, S.J., Moore, S.D., and Siggins, G.R., Ethanol augments GABAergic transmission in the central amygdala via CRFI receptors, *Science*, 303, 1512–1514, 2004.

NPG 8715.3, *NASA Safety Manual*, National Aeronautics and Space Administration, Washington, DC, January 24, 2000.

NSPE, The engineer will have proper regard for the safety, health, and welfare of the public in the performance of his professional duties, Section 2, *Code of Ethics for Engineers*, NSPE Publication 1102, National Society of Professional Engineers, Washington, DC, July 1976. (Also contained in *Ethics for Engineers*, NSPE Publication 1105, February 1976.)

Oberg, E. and Jones, F., *Machinery's Handbook*, 19th ed., Industrial Press, New York, 1973.

OHSAS Research Report 280, *Real Time Evaluation of Health and Safety Management in the National Health Service*, prepared by the Occupational Health and Safety Advisory Service (OHSAS) and the University of Aberdeen for the Health and Safety Executive, 2004.

Operational Risk Management: Guidelines and Tools, Air Force Instruction 90-901, Washington, DC, 2000.

Panksepp, J., Feeling the pain of social loss, *Science*, 302, 237–239, 2003.

Peters, G.A., *Human Error Principles*, Report ROM 3181-1001, Rocketdyne, a Division of North American Aviation, Canoga Park, CA, March 1, 1963. (Also in *Missile System Safety: An Evaluation of System Test Data*, Rocketdyne Report R-5135 (NASA N 63-15092), and reprinted in *AFSC Safety Newsletter*, August–September 1963, pp. 14–16, and *Approach* magazine, May 1994.)

Peters, G.A., Product liability and safety, in *The CRC Handbook of Mechanical Engineering*, Kreith, F., Ed., CRC Press, Boca Raton, FL, 1998, pp. 20-11–20-15. (Second edition in press.)

Peters, G.A., Hall, F., and Mitchell, C., *Human Performance in the Atlas Engine Maintenance Area*, ROM 2181-1002, Rocketdyne, a Division of North American Aviation, Canoga Park, CA, 1962.

Peters, G.A. and Peters, B.J., *Warnings, Instructions and Technical Communications*, Lawyers & Judges Publishing Co., Tucson, AZ, 1999.

Peters, G.A. and Peters, B.J., *Automotive Vehicle Safety*, Taylor & Francis, London, 2002a.

Peters, G.A. and Peters, B.J., The expanding scope of system safety, *Journal of System Safety*, 38, 12–16, 2002b.

Peters, G.A. and Peters, B.J., Design safety compromises, *Journal of System Safety*, 40, 6, 26–29, 2004.

Peters, G.A. and Peters, B.J., Legal issues in occupational ergonomics, in *Occupational Ergonomics Handbook*, 2nd ed., Marras, W.S. and Karwowski, W., Eds., CRC Press, Boca Raton, FL, 2005.

Pizzi, R.A., Anniversary of a Tragedy, *Today's Chemist at Work*, November 2004, pp. 29–31.

Price, B., *Chemical Processing*, February 2005, pp. 22–25.

PRS, Cannabis use impairs development, *Science*, 309, 222, 2005. (From *Proceedings of the National Academy of Science, United States of America*, 102, 9388, 2005.)

PUWER, *Provision and Use of Work Equipment Regulations* (U.K.), European Union, London, 1993.

Quality System Regulation, *Code of Federal Regulations* (U.S.), Part 820.30.

Raheja, D., *Assurance Technologies: Principles and Practices*, McGraw-Hill, New York, 1991.

R.A.K., Designing the rules is not always enough, *Science*, 209, 1980.

Rechtin, M., Sludged, *Automotive News*, April 18, 2005, pp. 25–28.

Reid, R.D., FMEA: Something Old, Something New, *Quality Progress*, April 2005, pp. 90–93.

Robinson, T.E., Addicted rats, *Science*, 305, 951–953, 2004.

Roethlesberger, F.I. and Dickson, W.J., *Management and the Worker*, Harvard University Press, Cambridge, MA, 1939.

Rotunda, R.D. and Dzien Kowsk, J.S., *Legal Ethics: The Lawyer's Deskbook on Professional Responsibility, 2005-2006 Ed.*, Thomson West, Egan, MI, 2005.

Ryan, M.A., Living with Lactose Intolerance, *Today's Chemist at Work*, December 2004, pp. 59–61.

Sanders, M.S. and McCormick, E.J., *Human Factors in Engineering and Design*, 7th ed., McGraw-Hill, New York, 1997.

SHP, Offshore Contractor's Injuries Were Caused by Wider Management Shortcomings, *The Safety and Health Practitioner*, October 2003, p. 8.

SNAME 4-22, *Sample Model Specification for Human Engineering Purposes*, Technical and Research Bulletin 22, Society of Naval Architects and Marine Engineers, Jersey City, NJ, 1988.

Snyder, R.G., Schneider, L.W., Owings, C.L., Reynolds, H.M., Golomb, D.H., and Schork, M.A., *Anthropometry of Infants, Children, and Youths to Age 18 for Product Safety Design*, SP-450, Society of Automotive Engineers, Warrendale, PA, 1977.

Supply of Machinery (Safety) Regulations (U.K.), Statutory Instruments, 1993, No. 3073.

Swain, A.D. and Guttmann, H.E., *Handbook of Human Reliability Analysis with Emphasis on Nuclear Power Plant Applications*, NUREG/CR-1278, U.S. Nuclear Regulatory Commission, Washington, DC, 1980.

Trials of War Criminals before the Nuremberg Military Tribunals under Control Council Law No. 10, Vol. 2, October 1946–April 1949, U.S. Government Printing Office, Washington DC, 1947, pp. 181–183.

Truett, R., Big 3 play catch-up in the hybrid game, *Automotive News Insight*, April 11, 2005, pp. 28N–29N.

USGS and SCEC, The magnitude 6.7 Northridge, California, earthquake of 17 January 1994, *Science*, 266, 389–395, 1994.

Van Cott, H.P. and Kinkade, R.G., *Human Engineering Guide to Equipment Design*, rev. ed., U.S. Government Printing Office, Washington, DC, 1972 (a revision of the 1963 *Human Engineering Guide to Equipment Design*, published by McGraw-Hill Co.).

Vanderschuren, L.J.M.J. and Eveitt, B.J., Drug seeking becomes compulsive after prolonged cocaine self-administration, *Science*, 305, 1017–1019, 2004.

Van Houten, B., Raising the bar, *Pharmaceutical and Medical Packaging News*, 13, 49–66, 2005.

VPP, Voluntary Protection Program, U.S. OSHA, 1982.

Wager, T., Rilling, J.K., Smith, E.E., Sokolik, A., Casey, K.L., Davidson, R.J., Kosslyn, S.M., Rose, R.M., and Cohen, J.D., Placebo-induced changes in fMRI and the anticipation and experience of pain, *Science*, 303, 1162–1167, 2004.

Waldman, P., Common Industrial Chemicals in Tiny Doses Raise Health Issue, *The Wall Street Journal*, July 25, 2005, pp. 1 and A12.

Watson, D., Crosbie, M.J., and Callender, J.H., *Time-Saver Standards for Architectural Design Data*, 7th ed., McGraw-Hill, New York, 1997, Appendix 3.

Woodson, W.E. and Conover, D.W., *Human Engineering Guide for Equipment Designers*, 2nd ed., University of California Press, Berkeley, CA, 1964 (a revision of the 1957 edition).

World Medical Association, *Declaration of Helsinki, 1964, Ethical Principles for Medical Research Involving Human Subjects*, June 1964, amended 1975, 1983, 1989, 1996, 2000, 2002, 2004.

Index